Chateau Musar

ACADEMIE DU VIN LIBRARY

Published 2020 by Académie du Vin Library Ltd
academieduvinlibrary.com

Publisher: Simon McMurtrie
Editor in Chief: Susan Keevil
Designer: Martin Preston
ISBN: 978-1-913141-04-2
Printed and bound in Italy by Elcograf S.p.A

CHATEAU MUSAR

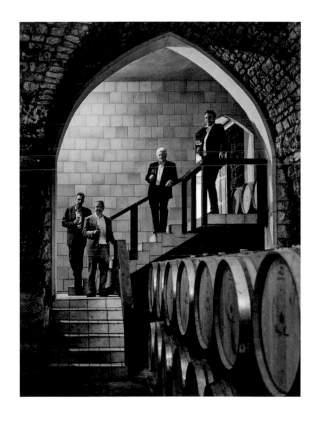

Edited by Susan Keevil

Photography by Lucy Pope

ACADEMIE DU VIN LIBRARY

Contributors

The publishers and the Hochar family would like to thank the following people for their valuable contributions to this book:

Bartholomew Broadbent
Michael Broadbent
Elizabeth Gilbert
Kevin Gould
Andrew Jefford
Susan Keevil
Sarah Kemp
Catherine Miles
Edward Ragg MW
Jancis Robinson MW
Steven Spurrier
Fongyee Walker MW

In loving memory of
Serge Hochar
winemaker (1939–2014)

Contents

Preface

SERGE HOCHAR (1939–2014)

You have to trust the messages that you are given. I trust the messages I get from my wines absolutely. Some of my winemaking decisions appear insane. I do not know how I came to them. Let's say that all the information on every subject is in the air, in the sky above you. Intuition is your ability to pull this out of the air. I feel sometimes I can be in communication with the universe. I thought this was common to everybody, but it seems it is not. Amazing!

We have to pursue people who are on the same wavelength as us. This person in front of you; I know he has a brain and millions of thoughts. My intuition, my information comes out of the blue, and sometimes out of his millions and my millions of possible thoughts; my intuition chooses one. It has happened to me all my life.

SERGE HOCHAR

Foreword

KEVIN GOULD

Whenever we talk of modern Lebanon, the very large elephant in the very small room is conflict. One of the reasons that Serge Hochar came to prominence in the wine world was that he was not only making outstanding wines of character but was making them on the front line of war. War has been a fact of Levantine life ever since people lived here, and it remains the red-stained ink on the end of millions of news nibs. But this book is no more about war than it is about peace. Instead, it is a tour through the craggy grandeur of Lebanon's landscape to meet the personalities who have created the country's, and perhaps the world's, most charismatic wines.

When I started writing my contribution to this book, Serge asked me how long it would take to finish. And would I hurry up, please, as he was intending to die in 2014. Good as ever to his word, he died while swimming off the coast of Mexico, on New Year's Eve, 2014. His death, like his life, was energetic, dramatic and unforgettable. His body may no longer be with us, but his spirit will surely endure forever in the Hochar family's wines with their unmistakable, remarkable and inimitable brand of humanity.

Like the wines of Chateau Musar and the conversation of Serge Hochar, this book ebbs and flows between the past and the present. It hopes to surprise, delight and enchant you, of course. But as much as anything, its intent is to embody the twin virtues that Serge single-mindedly pursued for his wines. So join me, please, in raising our glasses of Chateau Musar to truth and to life.

Foreword

MICHAEL BROADBENT

Dynamic yet charming, always good company, plus a superb vigneron and winemaker, Serge Hochar put the wines of Lebanon on the map. It is wonderful to note that the family is keeping them there.

I first recall meeting him in the late 1970s at one of the annual Bristol Wine Fairs. Christie's wine department always took a prominent stand and I would look after it assisted by my wife, Daphne, my son Bartholomew and our patient Bedlington terrier. One day, a friend came over to our stand, saying: 'You must come with me to taste the most remarkable wine.' On a small counter, with three vintages of a wine I had never before heard of, was Chateau Musar. Here, for the first time, I met the producer: a small, energetic charmer named Serge Hochar.

The vintage that appealed to me most was the 1967, which I subsequently wrote about, glowingly, in my monthly column in *Decanter* magazine. Of course, Serge Hochar was delighted and invited my wife and me (without the dog) to visit him in Lebanon. We went with my son, Bartholomew. It was not long after the cessation of war, the Syrians sending shells over the hills. At the time of our visit, Syrian soldiers guarded the main road across the valley. During these wars it could be a risky business to pick the grapes and deliver them safely to the Musar cellars outside Beirut. Thank goodness, things had settled down and the Hochar family could continue making wine.

My most interesting experience, and a new one for me, was the very personal way in which Serge tasted the young wine, noting its charac-

ter and its potential, and making the selection of grapes for his final blend.

I recall some of his early overseas buyers complaining they found the variation of styles confusing. They expected uniformity. Serge resisted these comments, explaining the reasons behind the vintage variations and how these made Musar so distinctive and appealing.

What is the wine like? I have a glass of the red 2008 vintage in my hand.

First, though it is clearly a red wine, it has a different hue, for example, than the more familiar (to me) Bordeaux. It has a deep, rich colour, ruddy, rather red, opaque core and a mature brownish rim; its bouquet is unlike any classic burgundy or Bordeaux, fragrant, almost dense; – on the palate, the distinctly sweet 'entry' is

Deep in the cellars, the red wines for which the Hochar family are famous slowly develop their unique character.

packed with fruit; it is a substantial wine, 14%, which motivates its sweetness. A rare, rich and harmonious red that I can drink by itself, and passes a great test of balance. Nothing juts out.

Having opened it I shall drink it tonight, thinking about Serge and the Hochar family. I will think how, in the face of difficulties that would cause most producers to give up or go bust, this family continues to make beautiful, distinctive wines in a beautiful if troubled land.

As for Serge: well, for all of us who had the privilege and pleasure of meeting the man himself, it is that very variation on a theme that makes his life's work so memorable.

FAMILY BUSINESS

How Chateau Musar came to be

Chapter 1

S E R G E - I S M

We can only trust these wines to be themselves. But what they will become
would surprise even the grapes.

SERGE HOCHAR

HOW TO MAKE RATIONAL the Musar magic? Is the spell it weaves spun by the wine itself, or by the dynamism of the man who created it? Deep weeks and long months have been spent learning to understand the wines in this book. They've been tasted in sumptuous apartments and cold bare rooms, first thing in the morning, last thing at night; with simple water crackers and with strong Lebanese food; tasted when happy, tasted when bored; in company and alone; tasted abroad and tasted at home. Yet always the magic has been there. At times they have been quirky, but they are always complex and distinctive. Musar wines are always unique.

And the winemaker himself, Serge Hochar? His wizardry and way with words has long been renowned. His estate, Chateau Musar, features on the world's finest restaurant lists and has a cult following through every wine loving nation. Yet the loyalty it inspires comes not from the flavours of clever oenological trickery but from a wine that is created by doing as little as possible, letting nature take its course.

Serge's triumph as a winemaker has not been through gimmickry and slick winery technique but lies in the all-consuming passion with which he believes in his wines and the life that moves within them. His single-minded determination to generate worldwide understanding of a wine that can be changeable, eccentric and unquestionably beautiful, has quietly catapulted Musar to global stardom. The effort that it has taken to do this – in the face of bombing, shelling, blockades and downright disbelief – has been almost superhuman. As with his wines, Serge's uncommon charisma and unstoppable energy go unmatched.

This is the story of Chateau Musar's rise to icon status and beyond. In the 30 years following Serge Hochar's decoration as *Decanter* magazine's Man of the Year in 1984, his wine became legendary. Unlike other high-status wines in this style (Right Bank Bordeaux or Côte d'Or burgundy, for example), there was no fabulous inherited property from which Musar could draw fame, no legacy of extraordinary wines with a triumphant set of awards and a centuries-old story to tell. Yes, Musar had an ancient *terroir* with a multi-millennia grape-growing performance record on which to plant its vineyards. But the winery's history began in 1930, only 90 years ago; in wine terms, it is a comparative baby. And yet the legacy carried forward by the Hochar family is huge...

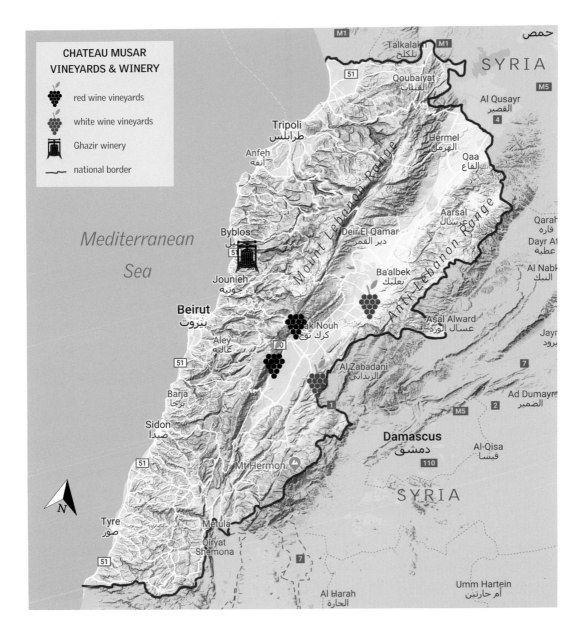

CHATEAU MUSAR VINEYARDS & WINERY

- 🍇 red wine vineyards
- 🍇 white wine vineyards
- 🏛 Ghazir winery
- ∿ national border

Mediterranean Sea

SYRIA

Tripoli
طرابلس

Anfeh
أنفه

Byblos
جبيل

Jounieh
جونيه

Beirut
بيروت

Aley
عاليه

Barja
برجا

Sidon
صيدا

Mt Hermon

Tyre
صور

Metula

Qiryat Shemona

Talkalakh
تلكلخ

Qoubaiyat
القبيات

Al Qusayr
القصير

Hermel
الهرمل

Qaa
القاع

Deir El Qamar
دير القمر

Aarsal
عرسال

Ba'albek
بعلبك

Mount Lebanon Range

Antx-Lebanon Range

Qarah
قاره

Dayr At
عطية

Al Nabk
النبك

Asal Alward
عسال الورد

Jayr
رود

Krak Nouh
كرك نوح

Al Zabadani
الزبداني

Ad Dumayr
الضمير

Damascus
دمشق

Al-Qisa
قيسا

SYRIA

Al Harah
الحارة

Umm Hartein
أم حارتين

حمص

Setting the scene

From its mountainside vantage point, Chateau Musar casts breathtaking views far out into the sparkling Mediterranean – the same views that Phoenician ancestors would have admired 3,000 years ago as they prepared to set sail, carrying wine and timber to trading partners across the sea. The strong stone buildings here are the beating heart of the property. Situated in the peaceful Silk Road village of Ghazir, this is the perfect refuge for making wartime wine – 25 kilometres beyond the bustle and bombing of Beirut; 100 kilometres from the battle-zone vineyards of the Beka'a Valley.

Ghazir might be the place of greatest safety, but historically, the Beka'a has always been the

The stunning view from the Hochars' old winery at Ghazir, perched on the mountainside overlooking Beirut.

main option when it comes to establishing vineyards in Lebanon. A flat, fertile plain nestled between the country's high Mount Lebanon and Anti-Lebanon mountain ranges, it is a colourful landscape – beautiful today with its tapestry of agricultures woven with tomato vines, olive groves, maize borders, fruit and cotton, but not always. Beka'a marks the Lebanon-Syria border zone – the battle scene for the violent Syria-Shia-Sunni skirmishes of the Civil War – and it is no place to build a winery and safely store treasured casks of maturing wine. It may be peaceful and productive now, but nobody knows how long the ceasefire will last...

Enter the essential 100-kilometre journey from vineyard to winery that is so integral to the story of Chateau Musar. Whichever way you look at it, it is this journey that has given Musar wines their extraordinary identity. It has involved danger, bravery and stalwart determination in times of war. In more recent peaceful times, it has required patience and forbearance. But without this transfer from war zone to coastal safe haven it is unlikely that the red or the white wines would exist today in the way they do.

The man who discovered the Ghazir-Beka'a formula, who had the inspiration to start it all in the first place, making wine from a string of villages in the Beka'a and transporting the grapes to the safety of the coast, was Gaston Hochar, Serge's father and founder of the winery.

Gaston was a visionary. As a young man, he went to study medicine in Paris, and while he was there fell in love with wine. The family banking business, which had grown rich in Ottoman lira by trading with India and the Far

East, lost everything when the empire collapsed in 1918. Gaston's only choice was to create a new path for himself, but the life of a junior doctor was not to be. (A generation later the same would happen to his son: once the desire to make wine had taken hold, there was no denying it.) In 1930 Gaston made the decision to return to Lebanon after his studies in France and begin a brand new venture making wine the way the French did.

The Hochar family was uneasy; wine – alcohol – did not constitute the 'acceptable' career path they would have chosen for him. But Gaston (tall, proud, good-looking and notoriously unswervable) had a challenging spirit and knew that the time was right for a new wine in Lebanon, and whatever he made would face little competition. The Jesuit monks, who had enjoyed a virtual monopoly of Lebanon's wine trade since the mid-1850s, made bulk-produced, ordinary wine that had little to do with Gaston's ambitions. He was determined to demonstrate the quality that he knew Lebanon was capable of.

As it turned out, Gaston had a ready market for his new wine. His first vintage was made in 1933, a time when Lebanon was still occupied by the French army instilled to govern the country after World War I. Eventually the French army bought the majority of the winery's production each year to be consumed by the army's officer cadre throughout the Levant. Business blossomed. Later on, in the early 1940s, Gaston befriended Major Ronald Barton of the Médoc Second Growth Château Léoville-Barton who was serving in Lebanon and encouraged him in his winemaking endeavours.

The next generation

In 1938, Gaston Hochar married Marcelle Achou. An energetic and driven woman who founded the nursing school of the Lebanese Red Cross, she had a quick mind and a sharp tongue. In Lebanon, the 'land of the show-off', she was uninterested in visible wealth and wore no jewelry. She refused to recognize the traditional subordinate place of a woman in the Lebanese home ('I am not here to keep house and wipe kids' backsides!') and quickly took up the role of adviser and financial controller to Gaston's wine business, encouraging him to expand his vineyard holdings and plant new vines.

Serge would say: 'My mother had anaemia when she was expecting me, so her doctor prescribed her a glass of red wine every day. We are addicted to doctors here in the Middle East, so she did as she was told. Perhaps that's where it all started?' And he'd continue: 'Mine was a reluctant birth; I was many days late. Eventually the doctors used forceps to pull me

Gaston Hochar Sr, founder of the dynasty that would go on to create Lebanon's most iconic wines.

out: hence my crazy brain. Perhaps I was happier in there sharing that glass of wine a day!'

Serge, Gaston's first son, was born in 1939, and Ronald arrived in 1944. The boys grew up in Beirut with their three sisters, and, following the family's Catholic tradition, were educated by the monks of the city's Jesuit school.

The Hochars are thought to be descendants of a Norman knight who followed the Crusaders to the Holy Land in the 11th or 12th century. The knight Hochar made his home on Mount Lebanon in a community that became followers of Saint Maron, who was venerated for the inspiration he took from the physical world and the deep faith it gave him. The Hochars are still members of the Christian Maronite Church – as are many who live in this part of Mount Lebanon, and around 25% of the country's population.

Serge and Ronald were regular visitors to the winery in Ghazir during their childhood, where they were encouraged to lend a hand, taste the wines and enjoy the buzz of the winemaking process. But Gaston advised them to follow their instincts (rather than himself) in their career paths, and always encouraged them

to strive to reach the top; he'd say: 'If you want to be a shoeshiner, be the best shoeshiner.'

As they grew up, Serge determined to study civil engineering, and Ronald law and music. But it soon became obvious that Serge's real passion was not for building design but for wine: he began to call it his 'reason for living'. The Hochars realized that if Serge were to follow his father in the business, his creative but tempestuous nature would need the steadying influence of his rational younger brother alongside. Much to the relief of his father ('We will lose him. He will end up being a piano player in a bar!'), Ronald gave up his place at the Conservatoire in Paris and followed Marcelle's example in becoming the steady financial hand in the business.

And Serge? Serge followed his dreams and decided to study winemaking at the University of Bordeaux under the tutelage of renowned professors Emile Peynaud and Jean Riberau-Gayon. He graduated in 1964, having taken an internship at Château Langoa-Barton, where

The young Ronald (left) and Serge Hochar sample their father Gaston's wine.

he began to observe and to understand the complexities of the grapes he wanted to work with. As his ideas crystalized, the wines he made began to develop their own unique characters. It became clear that in the winery Serge was an artist, and more than that, he had the stamina to make the wines he envisioned become reality.

By 1959 (even before he had graduated from wine school) Serge had developed enough confidence to take over the winemaking at Ghazir from his father, making red and white wines that were so fine, rich and complex that they were still wowing the world 30, 40 and 50 years later. This vintage, he said, was: 'The year that I was born, my way.'

From 1964 to 1970, Serge's wines were influenced by his teachings from Bordeaux, where he favoured the Saint-Julien appellation. After this time, he gradually shed his Bordeaux influence, and steadily developed the distinctive Lebanese product he would come to believe in so passionately. But the two brothers were being tested. The French army had left Lebanon in 1946 and Serge and Ronald were having to reach further and further into their own country for sales. With the onset of war in 1975, tourism and the trade that came with it dried up completely, and many Lebanese people moved away. For the first time, the Musar team was forced to find new markets outside their country.

The world takes notice...
Starting in London, in a small backroom at his father-in-law's travel agency on Sloane Street, Serge began to invite the world to share the secret of Musar. His first major breakthrough came at the 1979 Bristol Wine Fair in the UK, where Michael Broadbent was entranced by his wines, calling them 'the find of the fair'. He

told the wine world so in his regular *Decanter* column. The cognoscenti started to take notice.

Spurred on by the spirit of his Phoenician forebears, Serge then set out to find more routes to trade, and discovered his own gargantuan appetite for travel. Every few weeks he would either meet with sommeliers and wine lovers in the United States, conduct tastings in the Far East, travel to Brazil (with its Lebanese diaspora) to give wine lectures, or attend the grand wine fairs of Europe. The wines of Chateau Musar became better known abroad than they were in Lebanon, Serge earning himself a reputation not only for being a maker of beautiful, intriguing wines, but as a purveyor of elliptical measures of wine wisdom, or 'Serge-isms' (each chapter opens with one, and to experience a few more, *see* Chapter 11).

All would have gone smoothly but for the shelling in the Beka'a. Willing customers were one thing, but with the increasing damage caused by the war, there would be no wines to offer them. Musar was potentially in trouble. At home, through the battles, bombings and all-too-nearby skirmishes of the 1980s, Serge put in the most heroic vintages of his life, surmounting terrible odds to get his wines from the battle zone to comparative safety at the winery in Ghazir. His success under duress was rewarded by his commendation as *Decanter* magazine's first 'Man of the Year', a tribute[*]

---

[*] The *Decanter* 'Man of the Year' Award has been an annual tribute since 1984, received by luminaries such as Robert Mondavi (1989), Emile Peynaud (1990), Angelo Gaja (1991), Michael Broadbent (1993), Hugh Johnson (1995), Jancis Robinson MW (1999), Miguel Torres (2002), Christian Moueix of Château Petrus (2008), Aubert de Villaine of Domaine de la Romanée-Conti (2010) and Gerard Basset MW (2013). Serge is the 'benchmark by which successive recipients have been judged'.

that confirmed him as one of the wine world's most respected statesmen and set his wines on course for global acclaim.

That Serge enjoyed the success of his wines, there is no doubt. He made many, many friends on his travels, and was warmly welcomed by them each time he returned with a new vintage or new wine to try at a tasting, masterclass or grand winemaker dinner. 'Word of mouth, word of mouth, word of mouth,' was his mantra. The only way he felt his wines could fully be explained was by teaching in person. Nothing delighted him more than to see a glimmer of understanding, the beginnings of a recognition as a newcomer tried his wine. Then he would spring to life, crackling with intelligence, buzzing with temptations, and bamboozling with intriguing Serge-isms to develop that seed of Musar awareness: 'Stop! Don't taste it yet, get to know its aroma first – let its personality develop in the glass just a while longer, then take the teeniest, tiniest sip... Now you will see!'

Tragically, Serge Hochar died in 2014 in a swimming accident off the coast of Mexico. The family rarely had holidays and this Christmas trip was one they had been dreaming of taking together for years. In a sad twist of his incredible intuition, Serge had predicted that this would be the year his life would end. In typical Serge fashion – taking the most he could from every situation – he left it until the final day of the year, December 31st, to make his word true.

Taking Musar forwards

Ronald Hochar was true to his word, too. He had never left Serge alone to run the company. Today he is still there, heading up a close-knit team of Serge's sons, Gaston and Marc, his own son, Ralph, and viticulturalist and winemaker (Serge's 'spiritual third son') Tarek Sakr. In the story of Musar's wines, their heritage and future that follows, each member of the team has his own contribution to make – alongside the words of Serge himself. Here is an introduction to the Hochar family, the storytellers:

Dapper, silver-haired Ronald, now 76, is the elder statesman of Chateau Musar today. With his penchant for Saville Row tailoring and silk ties, he is a cultivated man with a witty energy and words that issue like a river running through rapids. His never-ending stream of anecdotes mingle and mix in a constant torrent of eddies and non-sequitors before eventually joining together as the river meets the sea and he makes his triumphant point. Ronald is the Musar patriarch; it is in him that you can occasionally see and hear a glimmer of his brother Serge. His passion for Chateau Musar and for his family is the same. 'We keep together because we have a common spirit,' he says: 'It is what my father and brother would have wanted and they would have been proud of us.'

Taking up his late brother's baton: Ronald Hochar.

Safe pair of hands: Serge's elder son Gaston.

Serge's elder son, Gaston (born in 1966), is Chateau Musar's managing director and head of winemaking. He lives in Beirut with his Swiss wife (Switzerland is their bolthole outside Lebanon just in case war returns or times get torrid) and has three grown-up children, Serge, Isabelle and Camille. When he is in Lebanon, he visits the winery every day. Gaston is the tallest of the Musar men and perhaps the most studious; his eyes have a kindly glint that belies a sense of humour running deep, staying far below the surface when first you meet him, but bubbling up while tasting at the winery or chatting at dinner. He is charming, quietly spoken and rather earnest, with the same beautiful Hochar manners and elegant deportment the team seems to share.

Where his father was untamed and unpredictable, Gaston is measured and cautious. He is utterly committed to securing the future of the business for his children, and for theirs. 'If you give Gaston 10, he will give you 10 back,' says Ronald. 'He is a safe pair of hands. If you give him 1,000 it's the same: he will give you 1,000 back.' Gaston's role today is an important one: he is the gatekeeper, the decision maker when it comes to the final blend of Musar's red and white *grands vins* and other wines. He is the orchestrator of Musar's brilliant new wine, Koraï. And it has been Gaston's decision to step up the status of Chateau Musar's prestigious white wine and release it in magnums. 'It is time the world took our white wine seriously!' he says.

Marc is Serge's second son, born in 1968. After a career in the financial sector, Marc came back to Musar to help manage the business and gradually followed in his father's footsteps by promoting Musar's wines to its international audience. His love for his family and

pride in the company reveal themselves time and again in the dedication he shows to his work and the passion with which he introduces and explains each Chateau Musar vintage.

Ronald congratulates himself on being the one who successfully persuaded Marc back into the fold. But Marc sees it differently: 'Gaston has a very low tolerance to risk, but I am a bit the reverse, a bit of a black sheep. My father didn't originally ask me to join the winery. He didn't want to impose on me and he knew we would have clashed. But after the financial crisis in 2008 things were different; sales had dropped and needed to be picked back up. I didn't want to come back and run the winery day to day, but I realized that changes were needed that weren't happening. It was as if there were a lot of ideas but they were all under a cap. I needed to help take off the cap!'

Marc has the same gentle twinkly eyed amusement at life as his father and brother, and has a flair for communicating his ideas with energy and clarity. But he is also intensely serious. Based in Munich, where he lives with his British-German wife and their daughter, Mia, Marc travels extensively to the Americas and the Far East, spreading news about Musar.

Ronald's son, Ralph Hochar, born 1974, is responsible for sales to France and Southeast Asian countries. Just as his cousins do – and his uncle did with such gusto – he enjoys the cosmopolitan life of a traveller, finding that many of Musar's customers throughout the world take such a keen interest in his visits and the new wines he brings that he now counts them as friends. Ralph also handles Musar's social media presence and while always smartly suited (just as all the Musar men are) he appears as much attached to his phone as any teenager would be – of course, this is a necessity of the job.

Ralph is thoughtful and sensitive. He says that of all new-generation Hochars, it is his daughter who most resembles her great-uncle Serge. 'She is a bit of a joker, just as he was. A comedian, a philosopher and a charmer. People were never indifferent to Serge; they loved him, but he could put them in their place too. He was complicated but he was genuine. He didn't try to be different in the way people do now on social media.' Ralph enjoys living in London with his Australian wife and their two children, Natalia and Matteo.

Black sheep: Marc Hochar returned to the fold in 2010.

Cousin Ralph (left) and 'spiritual third son' Tarek Sakr.

 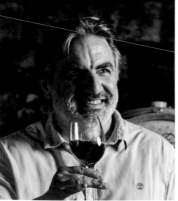

Tarek Sakr is the winemaker and viticultural-ist Serge always referred to as his 'spiritual son'. When he is not planting Musar vineyards, managing their cultivation, supervising the harvest or making wine, Tarek maintains his family's olive groves and admires fast cars (he would like to drive a 400-horsepower Porsche Cayenne S, but contents himself with a Subaru at 270hp, which is 'the nearest to love' he can get with a job in the vineyards). Tarek is archetypically Phoenician (his DNA proves it): broad shoul-dered and strong, with waving salt and pepper hair and piercing eyes that weigh up your worth in an instant. He is a talker, with deep-voiced, forceful views and effervescent creativity, yet he was happy to take orders from Serge.

Serge saw Tarek's potential for winemaking genius and took him under his wing in 1991. Serge said: 'Tarek has to follow my rules. When he came in 1991, I had established my ways, so instead of teaching him what to do, I told him what to do. He was open-minded enough to learn.' Tarek and Gaston have a great dialogue, constantly comparing ideas – though, accord-ing to Gaston, Tarek's adventurousness occa-sionally needs reining in. ✤

NATIONAL TREASURE

What makes Musar so special

Chapter 2

S E R G E - I S M

For me, 'identity' is a key word. Some people play poker and so do I, but the biggest gamble is to make wine. If you don't want to gamble, you have to kill it: stop it from being alive. In my wines there is a life you cannot stop. If you cannot stop it; it will live its own life. Fining removes those substances you strip if you want your wine to have less of a life.

The same is true with filtering – it removes some of the living organisms and reduces the potential for organic chemistry. These might be the nutrients that seduce your brain, that bring nuances to the wine. If each grape has 100 different dimensions, imagine how this is compounded in a wine! This is why Chateau Musar wines are so addictive.

The different dimensions communicate with your brain. The complexity of these dimensions increases during the two years each grape variety spends by itself before meeting other grapes.

SERGE HOCHAR

IT TAKES SUPREME confidence and unshakeable trust to make wine – any wine, but especially an iconic wine. The journey from grape to glass flows through numerous stages, during many of which a nervous winemaker will often conclude that either it will never be good, or otherwise that it will only be good if the winemaker intervenes. The Hochar way contrasts with this. Serge often insisted that he trusted his wines to make themselves, knowing that if he stuck to his customary back-seat approach in the winery, the wines, in the end, would all have their moment. He'd say: 'If a wine is not to your taste just now, then give it time; it will continue to evolve and it will often surprise you.'

Before they reach this 'moment', Chateau Musar's wines have a reputation for disappearing into a 'muted' phase, then emerging again some time – even some years – later, having shrugged off their teenage blues, to become

My Wines
are my Legacy.
When I have
finished talking,
they will talk
for me.

something tender, expressive and magnificent, the inspiration for extravagant tasting notes. At which point, Serge would shrug and say something along the lines of, 'I told you so'.

This sort of journey is by no means unknown in the world of grand wines. As such, Chateau Musar is often described as being a stylistic mixture of Bordeaux, Southern Rhône and burgundy: its Bordeaux-like structure gained from Cabernet Sauvignon; its Rhône-like flavour nuances of Carignan and Cinsault, and elegance and finesse (although not the character of its aroma) matching that of a great burgundy. Any of these great wines can – and frequently do – demonstrate similar 'dumb' phases.

So just as with a fine Bordeaux or burgundy, it is critical to let Musar wines – red, and equally importantly, the white – age before you uncork them. Let them mature first, then see

The 'Serge-ism' adorning the wall of the Ghazir barrel cellar rings true for the Hochar family and its visitors.

their real beauty emerge. This is why the Hochars invest so much time, money and space keeping their 'Chateau' wines for seven years in the cellar before releasing them for sale. They save their customers from drinking the wine when it simply isn't ready to leave the bottle.

And never assume that all bottles will be the same. Serge believed in finding the merits of each individual bottle of his wine rather than comparing it to others of the same vintage. He'd encourage his tasting companions to seek out the individuality in each and every glassful, saying: 'Give it time to breathe; let it gently get to know its new surroundings. Then taste what is in your glass. Truly taste it; taste it with all your senses awakened. Taste with your eyes.

Let your tongue undress the wine very slowly. Taste as your tastebuds are tasting.'

The present digital age married to our era of mass production means that a handmade, small-production, quirks-and-all, counter-intuitive wine like Musar has become something of a counter-culture. The suspense as the wine waits in your cellar, and the 'not knowing when' a bottle should be opened, each add to the mystique surrounding the wine. The fact that its best incarnation may take up to 30 years to arrive, that it will evolve in the glass and change from one minute to the next, that there is often a significant variation between bottles of the same vintage, and that rather than taste a wine from one year, you must taste and appraise each individual bottle on its own merits, means it can be a tricky wine to get to know.

And musty bottles are by no means unknown. Yet even when a bottle exhibits what those in the wine trade understand as a fault, magic is still undoubtedly present in its textures, tastes and aromas – open the same bottle in five years and the 'faults' will very often have gone, with a wine of shear magnificence – a phoenix having emerged in their place.

The beauty of this wine, what makes it so thrilling to drink, and what conveys its iconic status, is its absolute perfection when its time is right. In this respect it is not unlike a great burgundy. At its best, it cannot be bettered. But to understand it, one needs patience.

If all this seems like a set of smoke and mirrors for obscuring imperfect wine until the perfect sample just happens to turn up, then read on. The following tasting notes from wine world professionals prove otherwise. They relay the real elegance of Musar, and the way this iconic red tastes at its peak of perfection. They show that the wait is worthwhile.

1980 'Pale rust red. Very vigorous, palate-grabbing start. Full of life with light tannins on the end. Very impressive, even if it's still quite youthful. Really very difficult to assign a score to. Wonderfully impressive in every respect and this tastes as though it has a great future. A note of decay dissipated. 14%. Drink 2008–30.'
Jancis Robinson MW, *Library Vintage Reviews*

1986 'Orange-tinged; sweet, chocolatey nose that opened up richly; full, ripe, Pomerol-style, with quite a bite.'
Michael Broadbent

1988 'Tasted at *Decanter* magazine's Millennium Dinner: very burgundian, perfect with duck. Sweet, soft and, yet again, delicious. An identical note made at my 75th birthday dinner at Vintners' Hall, London in May 2002 and three times since. The most perfectly mature red imaginable: *à point*. I am busily drinking it at its peak.' Five stars.
Michael Broadbent

1994 'Hot summer in the Baka'a Valley, 40°C in August and September. One can smell the singed grapes in the wine. Remarkably sweet. Lovely fruit.' Four stars.
Michael Broadbent

1999 'This is the basic palette from which Serge Hochar, no dry academician, creates his blends. First, a Cinsault from a single, soil-rich vineyard in Aana: deep core, young Provençal scent; delicious flavour, wonderful richness and flesh; soft tannins. Next, a Cinsault from Ammiq's very gravelly soil: fragrant, lighter style, more charm. Then Carignan from a lighter soil in the Aana district: deep, velvety; spicy *garrigue* scent (I was reminded of Corbières in the Languedoc), flavour more aromatic. Lastly, Cabernet Sauvignon from Kefraya's rocky soil: very deep

purple; sweet, lovely crisp flavour, excellent tannins and acidity. End taste of violets and blackcurrants. In the Chateau Musar cellars, Dec 1999.'
Michael Broadbent

2003 'This is a ripe, sexy Musar. It retains its elegant mid-palate as well as lovely fruit and fine balance. There is a slightly sweet tinge to the finish, perhaps a touch of a high-toned demeanor. The lush texture and surprising richness make this a rather luscious wine by Musar standards. It is showing little but baby fat right now. Fresh and youthful, it is just a friendly puppy today. I'm not so sure it is in the winery's preferred style, but it should be a very enjoyable vintage for consumers. Showing not a trace of age, it has a lot of potential and a long life ahead.' 91 Points.
Neal Martin, Robert Parker's *Wine Advocate*

'It really was a thrill to see how beautifully these wines age. When tasted young, they can sometimes seem a tad coarse and unruly, but with a bit of bottle-age, the reds mellow into extremely serious, well-made wines... It was remarkable how long they lasted in opened bottles – even the oldest of them. The reds, blends of supposedly equal parts of Cabernet, Cinsault and Carignan, ended up tasting like particularly rich, spicy red Bordeaux after many years in bottle... In general the older Musar wines are real bargains – particularly vintages 1998 to 2008, all of them drinking extremely well now.'
Jancis Robinson MW, *JancisRobinson.com*

Understanding the Musar way

If listening to Serge teaches us anything, it is that tasting wine is neither a competition nor a comparative art. To be appreciated, a wine need not be marked out of 100 or be given a star rating. Or trigger in the analytical brain a bell that rings 'oak' or 'Chardonnay'. These things only get in the way of enjoyment. One should be more open-minded.

To explain this, there is an easy analogy with the travel industry. The difference between tourism and travel is that with the former you

Jancis Robinson MW, Michael Broadbent and Neal Martin: three of the most respected palates in the business and three great advocates for Chateau Musar.

transplant your physical and emotional comforts somewhere else, usually where the weather is better, and the amenities more ritzy. Tourism includes everything from package tours to all-inclusive resorts and seeks to insulate and protect visitors from the world outside, transporting them to an acceptable place hardly more foreign than their own. The vast majority of the world's wine is factory produced in this way; its grapes and flavours are constrained to conform to the winemakers' and marketeers' concept of what the consumer will find acceptable. Where a wine is 'exotic', it is tempered and tamed so as not to frighten the drinker. The result is that the tourist drinker seeks comfort in an ever-narrower band of flavours that must above all else be consistent, even when consistency is not always desirable.

Unlike the tourist, the traveller leaves her or his preconceptions at home and seeks out new experiences. By immersion in a place, in its smells and sights, its people and its food, the traveller invites and allows the creation of a world of possibilities, where events have the capacity to interest, excite, bore, shock, entertain, but always affect the traveller's life in some way. Travel requires curiosity, a spirit of adventure, and the capacity for surprise. The traveller does not want pre-packaged, controlled, instant experiences, preferring instead to sense the *genius loci,* or animating spirit of a place.

A tiny percentage of the world's wines – experts think less than 0.5% – are made by hand, using natural yeasts, with minimal chemical intervention. In the same way as, say, handthrown pottery varies from piece to piece, these wines can be variable in quality, but what they

Marc and Gaston Hochar in the courtyard of the old winery at Ghazir, where Serge once washed bottles.

have over 'machine-made' wines is the capacity to excite and enchant us, and to transport the taster to somewhere special.

Take time to talk...

It is this immersive quality that is so fascinating in Chateau Musar wines. They open an opportunity for conversation, not just about them, but with them. As Serge would say, for a true dialogue to exist, each conversationalist must respect the point of view of other, they should have something interesting to say, and they should have an appetite to listen. At a higher level deep, immersive conversation is also non-verbal, intuitive and even metaphysical.

So many wines, even those with fine, famous names, engage the taster only in a monologue. They bark, or they drone on; they tell you what's what and you either sign up or ship out. It is worth listening out for the others, subtly different, that engage you in conversation. The difference is in the winemaker: as with any cooking, a wine offers the maker the opportunity either to impose her or himself upon it by forcing the ingredients to become something the winemaker wants them to become, or alternatively to work with those ingredients to help them to be the best they can be.

There's nothing wrong with the first approach *per se*, but it is perhaps Victorian to assume that by dominating natural ingredients the man-made resulting whole will be greater than the sum of its parts. Subscribers to this school of cooking and winemaking seem to be intent on teaching their ingredients a damn good lesson. Their dishes and wines then teach that lesson to the consumer.

Fortunately for the many of us who are not naturally good students, there is an alternative approach, where winemakers or cooks see their work as encouraging their ingredients to express themselves to the fullest-possible extent, often by doing as little as possible. Chateau Musar, of course, falls assuredly into the second camp.

One of the most famous Serge-isms was heard when Serge told a room of wine writers that when making wine, he did: 'Nothing.'

'You do nothing Monsieur Hochar?' was the unbelieving response. 'Madame,' he replied, elegantly: 'We do nothing. But we do it in the right order.'

The difference between monologue and dialogue wines therefore most likely lies in the winemaker's ego and their willingness to put it temporarily to one side out of a dignified respect for his or her ingredients.

___

From vine (below) to bottle (right) – according to Serge Hochar, with very little intervention in between.

'Minimal intervention' is a wine industry buzz term. Some take this to mean farming without chemicals and (as in the biodynamic methods employed by some of the most celebrated Burgundian *marques*) ploughing with horses; then fermenting only with natural yeasts. But minimal intervention is also a state of mind as much as a way of working. The Hochar family have been practicing this for decades. They grow the very best fruit that their fields can produce using organic farming methods, then they treat these grapes with extreme kindness and respect. This kindness extends to giving the juice of that fruit the benefit of a great deal of time in order to allow it to become the fullest-possible expression of itself.

Because Musar wines embody the ethos of minimal intervention, because they are made with infinite respect to the grapes that gave them birth, and because they are not filtered – or 'lobotomized' as Serge would have it – they are famously long lived. Reds and whites from the 1959 vintage are today still fresh, alive, interesting to drink. (*See* Chapter 10).

So the winemaking facts of Chateau Musar laid bare are all very well (for their full details *see* Chapter 4), but they only hint at what the wine really is, and what it can become. The real magic comes from respecting and trusting the ingredients – the grapes from the Beka'a Valley, which are unparalleled in their character and intensity – and then allowing the wines to evolve and mature in their own way. Then, when they are ready, they will take you on a journey of the senses, conjouring memories, evoking emotions and connecting on many different levels. Combine these magical sensations with Musar's reputation as a plucky war wine from an off-piste location in tricky Lebanon, and the results become iconic. 🏵

THE BEKA'A

A *terroir* like no other

Chapter 3

S E R G E - I S M

This place is a blessing for the winemaker. In our wines you taste its sun, but not the heat. Even though we are further south than Europe, the altitude is high. This means you have the benefit of the sun, but without too much of its power to burn. Also, there are four distinct seasons. In winter, the Beka'a is white with snow like a duvet, so the vines can rest then wake up refreshed. In spring we have mild sunshine and rain: this brings new green life to the vines. The long, dry summer helps the grapes grow fat and juicy until we pick them by hand. Autumn is a little rainy to encourage the weeds to grow, which we just plough back into the earth in December to nourish the vines over winter. That is all.

SERGE HOCHAR

THE MAJORITY OF LEBANON'S grapes are grown in the Beka'a Valley, and for good reason. Not only is this vast agricultural region blessed with prized 'terra rossa' limestone soils – the holy grail of winemakers and viticulturalists the world over – but it is a full 10°C cooler than Lebanon's sultry coastal zone. The Beka'a is pretty much a paradise for grapes, but in the context of Musar it has one major complication: it is 100 kilometres away from the winery. And reaching it can at times be difficult.

Journey to the Beka'a
Leaving Beirut via its rambling eastern suburbs is easy enough today, give or take the odd army checkpoint, but as the route begins its 1,650-metre climb over Mount Lebanon it quickly

The Beka'a Valley looking east from Mount Lebanon.

steepens. Trucks and lorries in the adjoining lanes begin to growl and grind their gears as they struggle up the increasing gradients. It takes a while to break into open countryside.

The gently winding pavement over the mountain at first offers bleak viewing through the smog back to the city. To either side, the rolling foothills are punctuated occasionally by clusters of houses, or the gash and scree-slope of unpermitted open mine-work. Here and there, a terrace of green – home to apples, pears, vines and olive trees – is a verdant reminder that these dusty contours were once covered in cedar trees. But this Lebanese icon now appears in numbers too scattered to offer any reminiscence of the green carpet they once laid here.

And as the road rises, the full spread of Lebanon's countryside emerges: looming above and sprawling below, it becomes clear how

significantly life outside the city of Beirut is dominated by 'the mountain'.

The craggy bulk of Mount Lebanon takes up much of the central part of the country. So much so that during Ottoman rule, rather than living in Lebanon one was considered to live 'on The Mountain'. Topographically, today's Lebanon comprises two mountain ranges – Mount Lebanon, which runs north to south along the coastline, and the Anti-Lebanon range, which lies east of Mount Lebanon and marks the eastern border with Syria. Were Lebanon laid out flat like a tablecloth, it would be a sizeable country, but folded and creased as it is into mountains, passes and valleys, it takes up little space on the map. It's about half the size of Wales in the UK, or the size of Maryland in the US. The topography and road system being what they are, this can mean that somewhere a mere 15 kilometres away can take a whole morning to reach. But were you to drive up the coast, in clear traffic, you could get from Tyre in the far south of the country to Tripoli in the north between breakfast and lunchtime.

The road to the Beka'a climbs, dips, twists and climbs again, eventually reaching a high escarpment of 1,500 metres above sea level. To the left, Mount Lebanon tumbles messily down to the Mediterranean; down to the right, a plateau spreads suspended between this ridge and the broad bulk of Mount Hermon in the distance. The plateau, located at an altitude of 900 metres, is known as the Beka'a 'Valley'.

The Beka'a is actually not so much a valley as a long, flat, fertile plain that spans eight kilometres at its widest, and is hemmed in by the

Spring in the Beka'a, when the vines burst forth with green shoots following their winter slumbers.

two huge silent mountain ranges. From this height it looks like a bedspread sewn together from small, uneven patches of gold, terracotta, green and russet and flung neatly over the earth. (To gain some idea of how high 1,500 metres is, Ben Nevis, Great Britain's highest mountain, in Scotland, is just over 1,345 metres, while England's highest mountain, Scafell Pike, is not even 1,000 metres high. For

THE BEKA'A VALLEY AT SLUMBER AND IN SPRINGTIME

layers relatively quickly in rainy weather), but also reflects sunlight onto the vines, should they need any extra to help them ripen.

Striding out, arm extended in greeting is winemaker and viticulturalist Tarek Sakr. Tarek seems to crackle with a restless fiery energy, his hair the colour of iron and steel; his forearms Popeye-strong, and his chest broad and powerful. This is the man Serge introduced as his winemaker, his protégé and his spiritual son. Tarek joined Chateau Musar in 1991, aged 23, having studied oenology in France and trained at Château Lafite Rothschild.

Tarek admits: 'These boulders take a lot of clearing. It's heavy work. I know because I spent every day for 40 days planting vines here, each year between 2001 and 2005. I planted 45 hectares of vines, and without moving the stones, this old lake bed would be impossible to cultivate. And it wasn't only rocks; there were many tons of unexploded ordnance and spent shells too.'

Winemaker Tarek inspects his vines, gently watered in springtime by the melting snow from Mount Lebanon.

Sun, sun, frost and snow

The Beka'a is blessed with a great winemaking climate too. It would be easy to assume that Lebanon lies beyond the southern limit for viticulture. By latitude, it is 250 kilometres south of Sicily or the Strait of Gibraltar and some 1,200 kilometres south of Bordeaux. But its height (900m) above sea level allows it to be cooler than latitude would suggest, receiving up to 300 days of sunshine per year, but at a not-too-scorching average of 34°C in west Beka'a. And while frost isn't unheard of in the winter (in 2015 Chateau Musar lost 45% of its grape crop to frost in May), temperatures mostly hover above a kindly 10°C.

Being in the rain shadow of the mighty Mount Lebanon helps to keep Musar's vines thirsty, so they dig deep for water. 'You see vines, figs and olives growing together all the time, because they don't need much rain,' explains Tarek. 'Vineyards here in the northwest Beka'a receive an average of only 230mm of rain annually, where in the central valley, it's more likely to be 610mm.'

And do they irrigate? 'No, not at all. Irrigation is used for vegetables and fruit, but not for Musar vines,' emphasizes Gaston Hochar: 'We let the vines struggle so that their roots are forced to reach deep into the soil. They then have low yields but the grapes have increased taste and concentration.'

The well-fed drainage basins of the Beka'a's two main rivers make up for the lack of rain: the Orontes, flowing northbound into Syria then Turkey, and the Litani, flowing from the centre of the Beka'a south, then west into the Mediterranean, help distribute water around the valley. It all drains from the steep slopes of Mount Lebanon (3,088m), which force moisture-laden clouds from the Mediterranean to rise, cool and condense as they move east, thus releasing over a metre's worth of rain a year, which filters into the porous fractured mountain soils. The limestone of the foothills acts as a kind of aquifer, slowly releasing water to the vines, then refilled by rainfall each winter, and each spring by meltwaters from the snow. (Mount Lebanon is sometimes referred to as the Switzerland of the East, and can receive up to four metres of snow each year – which is welcomed by its six ski resorts.)

Vineyard villages

Musar's vineyards have historically been sited in a string of villages on the western side of the Beka'a Valley – Ammiq, Aana and Kefraya. They perch along a narrow band of limestone soils that flanks the mighty mountain, gazing out across the patchwork fields towards Mount Hermon. As foothills go, these slopes are short-lived, as the walls of Mount Lebanon quickly steepen to vertical behind each village, too sheer for any possible agriculture.

Life in the villages has been threaded and woven with grapevines since history has been written down. Both Aana and Kefraya are built around ancient, squat churches with vines espaliered onto their warm stone walls. Ammiq, where the family used to source grapes until the mid-2000s, is renowned for its natural wetland reserves (an important spot for bird-watchers) and the legend of Lady Shaawane, who dressed as a boy and escaped her life of luxury to devote herself to prayer.

Aana's vineyards are notable for their deep soils with patches of silt, clay and gravel over limestone, making the perfect base for the Cinsault and Grenache grapes that go into the Hochar Père & Fils blend. And Kefraya's vineyards for lighter stony soils over a base of limestone: these are the prize vineyards, perfect for Carignan, Cabernet Sauvignon and particularly for Cinsault. 'Wine from Kefraya has a certain elegance and finesse; it adds complexity to the blend, where Aana adds depth and structure,' says Gaston.

In addition to owning their own land, the Hochars also rent vineyards on the western side of the Beka'a. Marc Hochar explains that while owning these vineyards would be better, these things take time. The steepness of the valley walls means that the area of gently sloping land suitable for vines is relatively small and commands premium prices. Acquiring vineyards in the Beka'a takes very careful negotiation, patience and a great deal

The vineyard workers and their trusty shelter – necessary protection from the Beka'a Valley's searing summer heat.

of money. The family are lucky in having vineyard partners who are happy to work with Tarek and tend their vines in the Hochar way.

And within the villages, there are more villages. The Roman road, like a spine through the centre of the Beka'a, swishes past huddles of tents, trucks, makeshift houses and penned animals on its way. These are the homes of the semi-nomadic Bedouin people who live here in the valley from April until after the September harvest. Bedouin men and women make up the majority of the Hochars' grape pickers and they also work the land during the weeding and pruning seasons.

Since 2012 the Syrian Civil War has claimed many of the young men, who grabbed their guns and went to fight. By harvest 2014, over a million others had crossed the border and sought safety as refugees in Lebanon. (To put this figure in context, imagine if as a proportion of the UK's 66 million there were 15 million refugees and migrants in camps across the nation.) Suddenly, where one year there were almost no experienced pickers to help with the harvest, the next there were thousands.

But Serge would say: 'In Lebanon, difficulties are our habit. We are addicted to difficulties!' And he would find work for as many men and women as he could.

How the vines grow

The striking thing about the Hochar vines is that they at first seem to be grown in exactly the way they were back in Phoenician times, some 6,000 years ago. Yes, there are pergola-trained vines growing grapes in the villages, and yes, you'll see vines trained conventionally

---

A tool for the summertime: the hoe is vital for turning-in weeds and grasses so that they add nutrients to the soil.

on wires lining the roadside. But these are not Hochar vines. Hochar vines are allowed the freedom to grow as bush vines.

'When you train the vine on a wire,' says Tarek, 'you don't give nature the upper hand on the balance of its fruit and leaves. You need to let the vine translate the *millésime* [the vintage] to the grapes. If the vines are left on a trellis, the sun will take over. It's the same when you irrigate: you are not letting the vine tell you about the soil. So we don't irrigate. The vines must dig deep and find moisture themselves.'

Tarek talks at high speed, loudly and forcefully. You get the feeling that his vines (if they know what's good for them) will grow as they're told to grow. They certainly thrive in his care, looking abundant and healthy. Tarek explains that his bush vines are not left totally to their own devices. Pruning, which takes place from December to early April, is severe, cutting the vine back to five short branches, each with three or four buds. 'They need to work hard to grow!' By June, the branches will have reached out and upwards to form a small, dense 'tree', its foliage perfectly positioned so that the new grapes can receive both shade from the Beka'a's blazing sun, and enough light to ripen.

The vineyards are organic, as told by the heady hum of bee-life and constant chirrup of birds – audible only once Tarek's explanations pause. The sound of life buzzing, and the mix of wild oats and herbs that springs up between the vines (later ploughed back into the soil to act as a fertilizer and moisture trap) all add to the sense of enormous vinous well-being.

Clumps of pine and fir trees offer shade (the Cabernet Sauvignon vines need it) and protection from the wind that blows through the Aana vineyards: 'The vines are like the dress, and the trees are the earrings, or the

bright bonhomie of roadside fruit stalls create a sense of tranquilty that makes you forget the troubles once you reach the Beka'a valley floor.

'It's safe, until it's not safe,' says Marc Hochar with world-weariness. And somehow, sadly, the image of fighter jets slicing through the peace from one end of the valley to the other isn't hard to conjour as one considers how exposed the family vineyards in Ammiq, Aana and Kefraya were during the 1970s and 1980s. Wide, unfenced and as open to gunfire as they are to sunshine, the vineyards became part of the western side of the mountain controlled mostly by the Druze Militia led by Kamal Jumblatt, and by his son Walid after his father's assassination. The alliances between the Druze, the Christian Phalangists and various shades of Palestinian, shifted often. Opposing them, the Syrian army in the east was dug into Mount Hermon and into some parts of the Beka'a itself. The Israelis commanded the scene from the Golan Heights. The one stage that all three actors in this war had in full view was the Hochar vineyards.

Times were difficult, but Serge Hochar famously found every positive he could: 'Our vineyards were a shooting gallery, but some years the opposing armies called a ceasefire for our harvest. Sometimes they even came down

Gaston, Tarek and Marc inspect the Hochars' untrained bush vines as they thrive in the Beka'a's June sunshine.

to help us, especially if they knew that the ceasefire was not going to hold. Enemy soldiers would pick the grapes together before going back to prepare for the next battle,' recounted Serge, as though describing the impromptu 1914 Christmas Truce between British and German troops. 'Some years we just had to hold our breath and rush to get in as much of the harvest as we could. We were very fortunate that there was only one year during the wars when we did not make a vintage.'

Today the Beka'a lives not to the disgruntled crump of artillery and the dry rattle of automatic gunfire, but to the ancient rhythms of the seasons. Bedouin boys graze flocks of biblical black goats on fields of golden wheat stubble; pack mules are driven by men in baggy salvar trousers and untidy turbans, and sunny stacks of precariously balanced pumpkins still adorn the roadside. Were it not for the telegraph wires strung above the fence-lines and the roar of passing vehicles, it would be hard to say which century they were in. This high plateau, slung as it is between two mountains and two countries, is suspended in time, too. 🏵

Chapter 4

NATURE'S WAY
Winemaking in Ghazir

S E R G E - I S M

You and I can both make a wine at the same time, from the same grapes, in the same place, using the same recipe and techniques and the same equipment. Yet the wines will be different. They will have different souls, and will lead different lives. This is because wine has the ability and the capacity to understand the person who is taking care of it.

SERGE HOCHAR

At first sight, Serge Hochar did not conform to the romantic idea of a war-zone winemaker. He sported neither flowing, unruly hair, nor an extravagant hat or spectacles. His clothes were sober, verging on the dapper – a pair of expensive, neatly pressed slacks paired with a toning short-sleeved shirt, worn as if to blend in at the golf club bar. He had an aroma faintly of Old Spice. His clean-shaven features not dissimilar to those of ex-British Home Secretary Jack Straw. But where pragmatic Jack's eyes betrayed little of the real him, Serge Hochar's would burn like lava. This man, they'd say, has seen plenty yet knows no fear. Serge's look was at once challenging, disarming, frank, very bright and rather amused. His eyes warning, treat me with care, or else.

Serge was a legend. When the bombs were dropping on Beirut, when rocket-propelled grenades were exploding all around, and automatic gunfire cracked, pinged and thudded here, there and everywhere during 15 years of war, Serge stayed resolutely in Lebanon and got on with the business of making great wine.

He quietly picked his grapes, and when his winery was being bombed from the air and bazooka'd from the ground, he turned his cellars into shelters and threw parties.

It is Serge Hochar who made Musar what it is today. Who carved out the unique identity of this wine, and set a framework for those who follow him to stick by.

To the winery at Ghazir

To the left, the sea. To the right, the mountains. Of the five million souls in Lebanon, the majority live near its narrow coastal strip – four miles across at its very widest, often petering out to nothing. There's just about room for the six-lane coastal highway, a random splattering of shops and showrooms and a selection of luxury and by-the-hour hotels. Humanity on the move.

The drive from Musar's offices in Achrafiyeh, central Beirut, to the winery of Chateau Musar at Ghazir takes 20 minutes or two painful hours, depending on traffic. Follow Independence

The imposing limestone block exterior of the 'new' winery.

Road up over the hill, through Sassine Square and leave behind the world of the modern-day souk (Starbucks, Adidas and Victoria's Secret) and find yourself heading north.

A huddle of ancient American school buses repainted in the colours of the fairground lie waiting as the smart tide of Achrafiyeh Hummers, Maseratis and Porsches joins the flotsam and jetsam of the Middle Eastern road. Venerable Mercedes', their Frankenstein bodies patched and filled so many times that hardly any of the original car remains, squeeze into spaces they shouldn't. Pick-up trucks overfilled with watermelons, bales of waste paper or crated white goods – invariably driven by impassive, leather-faced chain-smokers – carry a complement of jeans-clad workers clinging grimly to the truck's tailgate. Grimey delivery motorbikes (helmet optional) dart between what has now become four lanes of furious hooting traffic.

It either takes a driver as ruthless as Serge was, or one of calm disdain, as his sons Gaston and Marc prove to be, to negotiate this turmoil. Vehicles are rarely allowed to pass; no driver may cut in front and get away with it; briefly vacant gaps are to be accelerated into. Indicators and rear view mirrors prove less essential than a frequently buzzing phone. To the Hochars, and to most Lebanese, driving is an extreme sport. And if it seems a touch hard on brake pads and passengers, no matter.

Just north of Beirut the seashore narrows so tightly that the Lebanese – in another hard battle – have struggled to retain it. A constant programme of reclamation is in progress. Travel north and the mountain pushes the road so close to the sea that the city suburbs, shops and squeezed-in apartments eventually yield to nature, as the limestone cliffs take over.

A cleft to the right marks Dog River, or Nahr el Kalb. For five millennia this rugged pass offered the least difficult route across the mountains to Damascus and beyond. For invading (and, inevitably, retreating) armies and generations of traders, this was the only way east. A proud collection of exhaust-blackened stellae, now preserved as a UNESCO world heritage site, marks the passage of every successive conqueror of Beirut for the last 3,000 years. Pharaoh Ramses II and King Nebuchadnezzar both passed this way with their troops, as did Greeks, Romans, Byzantines, Ottomans and French and British colonial powers.

The River Dog, running through this treacherous valley, was named after the piercing howl

The old winery's roof terraces benefit from spectacular sea views, making them ideal for family gatherings.

The medieval stone bridge built by the Sultan Makluk over the Nahr al Kalb ('Dog') river en route to Ghazir.

that can be heard when storm winds are forced through this narrow limestone gap. No doubt a similar howl to that felt by Serge when he and his grape trucks tried to negotiate this heavily guarded three-way pinch-point during the war. To get from the winery at Ghazir to the vineyards of the Beka'a and back again, all traffic had to pass this way, there was no 'route B'.

Nineteen kilometres later and the suburbs have barely abated. Although there's no money in real estate right now (war is still the deterrent), it's not for want of trying. Breeze-block piles seem to spring up in every nook. The road winds past the Casino du Liban at Jounieh (advertizing Vegas-style two cocktail bars, five restaurants, hundreds of slot machines and an extravaganza starring the extravagantly moustachio'd Tony Hanna with 'His Lovely Songs and his Charisma'). It then turns steep right and zig-zags (tyres squealing) into the painted stone village of Ghazir.

Ghazir was once a silk-making centre (as were many villages on The Mountain until the 1860s): a clumsy stone monument testifies to its place on Lebanon's *Route de la Soie* – the network of silk trading roads that linked east and west). High above Ghazir perches the Church of the Divine Ascension, whose imposing stone walls bear the bearded image of Jacob of Ghazir, the 'Apostle of Lebanon', a Capuchin Franciscan friar who lived here until his passing in 1954. Jacob's blissful gaze guards an improbably steep bend, at the blind apex of which there is a sharp turn left into the cobbled calm enclosure of the Chateau Musar winery.

The old...

Looking out to sea stands the old winery. Set slightly back, across the steep road, stands the 'new'. In 1930, on land that had been in the family for many years, Gaston Hochar founded Chateau Musar in a large, 18th-century convent possessed of a certain brutal glamour and

yard-thick, sand-coloured block walls. It sits on a grassy promontory that commands the coast road, with views straight out to the brightly turquoise Mediterranean 500 metres below and back through the mists to Beirut. It is soon to become a venue for expensive weddings and wine tastings, but right now, in its transitional state (glassless windows and stairways ending mysteriously in ladders), it welcomes with a hint of its future charm, the monastic grandeur of its past and the haunting but certain presence of olden vintages.

A tour round with Gaston, Marc, Ralph and Ronald gives an insight into Hochar winemaking history. Ronald remembers carts trundling into the courtyard laden with grapes. He remembers peering through leaded windows out to sea from the dark cocoon of the press room, grapes (and boy) safe from the searing heat outside. He remembers the squeak and crunch of wood turning and press stones crushing, then the splash of grape juice as it runs free into containers that stand ready to fill the vats.

Most of all, Ronald remembers running along wooden gantries above the concrete wine vats that line the press room, listening to the 'chug, chug, chug,' from below. 'The carbon dioxide bubbling through the fermenting wine was like the sound of a great engine; it had a rhythmic beat and you'd hear that glugging sound all day.' A row of nine vats lines one room, and a row of 10 once stood in the dimly lit room behind. Each would have been no more than 8,000 litres in capacity, some of them (as can be seen from the marks in the concrete where they once nestled) divided into smaller internal chambers so that Gaston (the elder) could ferment parcels of grapes separately according to site and variety.

The old stone fermenting vats of the original winery – and the gantry along which Ronald used to run as a boy.

But which variety, and which site? 'It's too late to know now!' laughs Gaston Junior mysteriously.

But Gaston Senior had an eye to making quality wine 'the French way', and his sons, Serge and Ronald, caught on quickly. It is likely that these small fermentation spaces held forerunners of the many experimental wines that led to the Chateau Musar we know today.

'My father wanted us to be involved but he didn't force us,' says Ronald. 'In 1957, when Serge was 17 and I was 14, we came to work here because the bottling machine wasn't behaving. We had to wash the bottles in the fountain in the middle of the courtyard. We had to wash 800 an hour, but the machine had done many times that. We couldn't go fast enough to keep up with demand for the wine. It wasn't like now when we give it seven years to mature before we release it, people wanted to drink it right away. They had to be patient then and wait for us. They have to be patient now and wait seven years!'

Standing in the courtyard now, surrounded by its arched colonnades, high entrance ways and wafting bougainvillea, it's hard to imagine the two teenagers and their inevitable banter. Serge and Ronald went to the Jesuit school in Beirut, each taught by the monks for 18 years. 'I trained in accountancy, the law and practical sciences. The Jesuits taught us how to think in a rational way, to put our passions aside. Then we came back here to work for my father.

'Back then when we were washing bottles, it was December and it was freezing! We stopped at 12 for a sandwich and finished at 4.30pm. Then we had to go back to Beirut to stick on labels. I used to earn five Lebanese pounds a day for this, so I could save up to buy a car when I was 18. (I bought a Peugeot 404.) Incentivization was his way and it worked. I could have been put off by that freezing cold and those long hours. And of course alcohol was ill-considered

at the time; it was not such a well-viewed product. But we were busy – we made 300,000 bottles a year, which sold immediately – and I came to love the wine.'

The path from the courtyard out across the stone paving and elegant lawns of the 'chateau' garden is flooded by Mediterranean sunshine. As he walks, Ronald admits: 'I also promised my father that I would not leave Serge on his own.'

...and the 'new'

Winemaking comes into sharper focus across the road at the Hochars' modern-day facility, built by Gaston Sr in 1958. Dug into the side of Mount Lebanon, the building is large and functional, constructed from sturdy, biscuit coloured cubes of limestone: four levels for the winery, and two at the top for the family's

Ronald, Marc, Ralph and Gaston Hochar arrive at the entrance to their Lebanese-styled Ghazir winery.

apartments (where they sheltered from the bombing in Beirut, 1975–1990). A vine winds itself around a fence beside the gate, an ancient olive tree casts its shade by the neatly cobbled entranceway, and bougainvillea attaches itself in magenta swags wherever it can. Before you reach the main door, there is a tiny chapel dedicated to Saint Rita, the kindly patroness of impossible causes and hopeless circumstances. Through troubled times, she has been very good to the Hochars.

The winery has a triple purpose, to make wine from grapes, to mature it in oak barrels, and to bottle and store it until it is deemed ready for sale at seven years old.

Straight from the Beka'a, the grapes arrive in truckloads at the winery door. They are then transferred from the receiving hoppers into the dark pit of the crusher destemmer where they are gently tumbled together in a process that removes their stems and stalks and starts to release their juice before they go to the vats for fermentation. Some wineries will leave stems on the grape bunches at this stage for extra tannin, but at Musar Gaston and Tarek follow Serge's principle that these tannins are too bitter.

Just as red and white grapes are both destemmed, pressing is the same for red and white grapes too. An enormous Bucher press has pride of place at the head of the winery. Sparkling clean, and waiting for its first load of grapes – likely to arrive as early as the first week in August, when the Chardonnay and Viognier will ripen in the Western Beka'a.

From the press, the juice is run into the gallery of vats that line Musar's impressive fermentation room. Just as Musar's wine remains unique in the world, so too does its winery. It is a 1970s build that the cousins seem ambivalent about today, but which perfectly matches the wine and its creator for unflinching individuality. Tiled in the colours of the Lebanese flag – red for the upper vats and the floor (on the flag this colour represents the blood shed by its people to protect their country), white for the lower vats, representing peace, purity and the snow that tops Mountain Lebanon – it is a concession to the art of wine (not to mention pride in their country) that few anywhere in the industry bother to make.

Not only are they stylish, the vats are made from concrete – another reflection of the era in which they were built, before the sensible craze for making wine in stainless steel took hold in the 1980s – which is very much the choice of today's winemaker. Concrete is a natural insulator and will stabilize the temperature of a wine. It is also porous and allows it to breathe and evolve where stainless steel, which allows no micro-oxygenation, can 'flatten' the flavour of a wine. Unlike oak, concrete is neutral, it doesn't impart flavour, but allows the characters of *terroir* to show through. This is another example of the Musar team being 'behind the curve', happy with their 1970s equipment, yet at the same time very much ahead of it in terms of modern-day wine thinking.

'Fermentation occurs naturally, with natural yeast, nothing added,' says Gaston. 'We have been making wine this way for over 60 years.' Musar vineyards are certified organic. This has nothing to do with marketing literature and everything to do with preserving the yeasts that naturally occur on the grapes. 'Any work we do in the vineyards, whether it's ploughing the soil or pruning the vines, will affect the yeast population, so we have to take great care. Each yeast strain is directly related

Winemaker Tarek Sakr (left) and Marc Hochar discuss the finer points of producing Lebanon's most famous wine.

to each part of the vineyard and to each grape, helping it to ferment in its own way, giving it its own unique character. It is really important that nothing disturbs this relationship.'

The red wines

'I trust my wines to be whatever they want to be,' Serge famously said. But for a few small decisions at the time of fermentation, trust is still a key part of the Hochar winemaking philosophy. The wines are very much left to their own devices. Natural yeasts, neutral concrete vats and low ambient temperatures enable a gentle transformation from juice to wine. It all happens at 20–26°C and takes seven to 10 days, depending on the will of the vintage.

But red grapes need to impart as much colour and flavour to their wine as possible, and to get this 'extraction' right the winemaker must intervene. Gaston needs to decide how long the skins and juice are left together to macerate, and how many times they are pumped over so that the floating cap of skins is refreshed and encouraged to release more character. 'It could be a week, three days or 10 days,' he says: 'It just depends on the year, the grape and the wine we are making.'

Gaston and Tarek taste the wine until it is 'dry' (all its sugars fermented out) then rack it back into a new vat to settle and undergo its second, softening malolactic fermentation. For the next nine months, the wine is once again left peacefully to its own devices. Then, it will make its way to the barrel cellar to sleep, undisturbed in oak...

'Voilà!' says Tarek: 'Wine is conceived in the vineyard, but it is here in the barrel room and

Tarek and Marc sample a future vintage – the final decision on when to bottle it rests with Gaston.

deeper down, in the cellar, that it grows up, and receives its education.' A quiet education. The atmosphere in the cellar is indeed studious. Under cathedral-high, vaulted brick ceilings supported by square brick columns sit row upon row of silent oak barrels, each marked with the elegant cursive logo of Chateau Musar. It is completely still here, the temperature mildly chilled: nearly enough for goosebumps, yet not quite cool enough for a jacket. It feels special, and slightly shivery.

Tarek explains that the barrels are made of Nevers French oak and around 10% will be new each year. 'With your children, you do your best to provide a school in which they can thrive,' he says. 'And the winemaker's job is the same. When the wine is made and put into these barrels, I am no longer the midwife, but must become its father, mother, teacher and friend.' Serge Hochar felt the same way about schooling the wine, he would say: 'Once you put wine in wood, there's contact with the air, both in racking the wine into the barrel, and through the fibres of the wood. This teaches the wine to face the life ahead, and it prepares it for a longer existence.'

The wine matures in oak for a year, it is then drawn out into concrete vats ready for blending. It is now two years old. Most wineries will blend their different grapes and wines a few short weeks or months after fermentation, but at Musar, Gaston wants to know the full character of each component first so he can gauge what it will contribute to the final blend. 'At two years it will have only just started its journey, it will still have many changes to go through. But we can know a little more clearly where it is heading, and the character it will have.'

The Hochars' finest wine, their *grand vin*, is 'Chateau Musar'. For this, Cinsault, Carignan

Maturation in oak plays a key role in the formation of the wine's character – Musar uses Nevers casks from France.

and Cabernet Sauvignon are the grapes used, from the vineyards of Kefraya. The parcel or plot of vines the grapes come from will be different for every vintage although the proportion of each grape in the final blend will be roughly thirds. Tarek admits that he and Gaston now know this wine so well that they can blend it almost instinctively: 'When I first joined the winery in 1991, it took four days to make a blend. Now it takes us four hours. Even from the harvest we have a very good idea of what we are going to use.' But the final combination can only be confirmed with the tasting of numerous samples from barrel and vat.

The Hochars make a series of introductory wines that lead up to their iconic *grand vin*, 'Chateau' as its affectionately known. There is 'Jeune', a deeply fruity, young-drinking red made from Cinsault, Syrah and Cabernet Sauvignon. It's made from young vines but also some older vines depending on the vintage, and released at two years of age: this wine is a celebration of the dark, velvety nature of Syrah. Then, Hochar Père & Fils is made from 30-year old Cinsault, Grenache and Cabernet vines, solely from the vineyards of Aana. It is aged for six to nine months in Nevers oak, and, as with the *grand vin* (which it resembles with its rich textures of fruit), is blended to reflect the character of the vintage. It is not released until two to three years of age. All the vineyards are grown to produce similar high-quality grapes with the same low yields, no matter whether they end up in 'Chateau', Hochar Père & Fils or Jeune. Finding the best balance of each wine during the blending phase is what determines the final allocation of a certain vat to a specific blend.

There are rosé wines, too, under the Jeune and 'Koraï' brands – each made from Mourvèdre and Cinsault. These are wines made from red grapes, but lightly pressed and fermented

Tarek tastes Carignan, mid-way through its year-long 'education' in oak barrel, after which it will be blended with Cinsault and Cabernet Sauvignon.

much more in white wine style. Jeune Rosé has 85% Cinsault and 15% Mourvèdre, while Koraï Rosé is 80% Mourvèdre and 20% Cinsault.

In the cellars

There are few other wineries in the world that dare (or can afford) to wait so long before releasing their wines to market. Most wineries need to sell their wines young to keep the finances healthy. But at Musar, time is a huge factor in shaping the eventual character of the wine; so much so that it will not be released until seven years of age, when its mature flavours show the very first signs of emerging.

Chateau Musar is said by many (the Hochars included) to be at its best from 12 to 15 years of age. To begin its journey to maturity the wine

The 'Musar Jeune' trio, white, red and rosé.

is bottled then stored in cellars dug deep into the moist limestone heart of Mount Lebanon – the lower two floors of the Ghazir Winery.

Last glance, a heat-haze view of the sea through a picture window, then the sliding metal grilles of a goods lift open. Nightclub dark, it clunks slowly down one floor, then another. The metal gates squeak-slide open to reveal a damp, dark gloom. Someone throws a switch, and an acre of cobwebby lightbulbs blink weakly on. Rib-high stacks of wine bottles, thousands upon thousands of them, fill the cellar from wall to hewn-stone wall. It is a vast subterranean sea of bottles, a mini ocean of fine wine.

This is the world's largest stock of a single wine *marque*: 70 vintages of one of the world's most famous wines. Some bottles are covered in a pencil-grey blanket of cobwebs, others are

mildewed, their labels almost unreadable. Bottles of the vintage white wines glow like bars of gold in a Bank of England vault. They lie just waiting and gradually changing. This is the underground kingdom of the late Serge Hochar, the Willy Wonka of wine, where his bottles develop their sensory magic. This is where his trusted vintages started to become what they want to be before people enjoy drinking them.

To the cool plink-plink drip of an underground mountain spring, the walkway through

---

The bottles are stored for seven years, their contents appraised at frequent intervals to check on its progress.

the cellar leads to a narrow wrought iron-gate so covered in black cobwebs it looks like part of the mountainside. Within, lie smaller dustier enclaves, filled with the real treasure of Musar, library vintages: wines dating back all the way to the 1930s when winemaking began here.

The white wines

Chateau Musar's white wine is unique. Serge Hochar was the first winemaker to recognize the blending possibilities of Lebanon's native Obaideh and Merwah grapes – indeed, from the 1930s to the early 2000s, the Hochars were alone in producing wine from these white varieties. With their distinctive characters they were sometimes hard to understand, but Serge saw their potential, and with them created a wine that became every bit as spectacular and ageworthy as the Chateau's red. As older vintages became available, tasters became intrigued by the idea of wines that were old (in terms of both Lebanese heritage, and bottle age) but new to their palate.

Today's wine drinkers, who generally expect to chill down the most recent vintage of a known white grape and drink it the same day, miss the point entirely if they do this with Musar Blanc. As with the red, it is a wine that needs time. When asked how for how long, Gaston talks of the 1954 which 'refused to grow old'. But find any wine over 10 years and prepare to be bowled-over.

As Serge said: 'My white is not understood. It is bigger than my red wine; it goes with more foods and it is more serious. But it is also more difficult. One day people will come to it.' And they should.

Serge Hochar always said that his white wine, a blend of Obaideh and Merwah grapes, was bigger than his red.

Tarek sees the Merwah and Obaideh grapes as growing boys: 'These two Lebanese varieties are completely different. Obaideh has lots of depth and a firm structure. The Merwah has pure, one-dimensional fruit and lots of nervous energy – it is so pure in that single dimension that it is almost overwhelming! When it comes to the finish Merwah still has this lemon-zest freshness and mountains of minerality. Alone it would always be fighting, always be striving, a man within a boy's body, not one ounce of fat! He needs more wisdom and age to calm down; someone to say: "You have to grow old young man, you have to be mature..." So Obaideh joins in: Obaideh is much more elegant and has structure. This lies down and lets the wine age. It dilutes the temper of Merwah. The wines from these grapes are practically like human beings and they are always in harmony with one another, and you always get an interesting result with them.'

In the winery, the grapes are destemmed, pressed, removed from the skins and transferred to oak barrels for fermentation (at 20–25°C). So far so usual. But where some winemakers might block the malolactic or secondary fermentation in order to retain a certain crispness and acidity in their wine, the Hochars believe in a more Burgundian approach, allowing the process to occur naturally, using the bacteria naturally present on the grapes (just as they do with the first fermentation) thus helping the wine to become richer and smoother. 'We do not block the malo because our phenological profile doesn't allow it. If we were to block malo we would have to use high levels of sulphur, which we are against,' says Gaston: 'We do not choose to control our wines in this way.' As with the reds, the evolution of Musar's white wines is left to Mother Nature.

The Musar Blanc blend comprises 60% Obaideh grapes and 40% Merwah, each barrel-fermented then aged for nine months in oak casks. The wines are blended within one year, then bottled and stored for seven in the mountainside cellars before they are released. As an example, 2010 was pale and unassuming in colour when tasted in 2019, but hit hard with a strong, mineral and orange-peel nose followed by smoke and peach kernels in a bold core of fruit on the palate. It packed a punch, yet still seemed reserved, waiting. The 2009, with an extra year's age, was full of apricot and orange blossom aromas, a gliding, unctuous richness braced by tangy acidity and a luxuriously peachy, complex finish.

To help newcomers to Musar's whites understand the amazing style and potential of these wines, the Hochars have developed Jeune Blanc, a younger-drinking wine, to lead up to it. It is a stunning blend of thirds each Vermentino, Viognier and Chardonnay. For this wine the grapes are destemmed, pressed and then fermented in stainless steel vats. The wine is released at six months old, with no oak ageing, but comes from very low yielding vines (18hl/ha) so its flavours are packed together, intense. Lemon citrus richness from the Vermentino mixes with peach and white pepper complexity from the Viognier, linked by an overarching familiar hazelnut smoothness from Chardonnay. 'Chardonnay is always the bridge that links Vermentino and Viognier' says Gaston, explaining: 'The secret of this wine is to have great Chardonnay; we have to choose the right harvest date, not too late or the wine will be too flabby.'

With the 2018 vintage, Musar has taken a step further towards 'Chateau Blanc' with the launch of its Koraï Blanc, which targets the local Lebanese market. This is 50% Viognier, 50% Vermentino, with 100% new oak for fermentation. This very fine, pale golden wine, starts with a delicate butter and melon nose, then ratchets up to fruity grapefruit-apricot intensity, the two grapes perfectly in harmony to a lingering finish surrounded by a fine, barely-there hint of oak. The presence of Chardonnay isn't missed.

These whites have the bare minimum of sulphur treatment and all have been through their malolactic fermentation; but unlike the reds (where a sediment is more acceptable) they are gently fined with clay before undergoing a brief cold stabilization. Hopefully these two younger white wines will show the way to a new appreciation for Chateau Musar Blanc.

'Koraï', Musar's new brand for the Lebanese market – lighter than the 'Chateau' wines, but more serious than the 'Jeune' range.

Serge's philosophy of slowness held that the longer a wine took to be made, the better it would age.

Musar and Time

For Serge and his wines, slowness was important. The Arab maxim 'If you don't cut through time, it will cut through you' was one he held dear*. And this is still the guiding principle behind Musar winemaking.

Serge acknowledged that everything in Lebanon has taken time to come to fruition – its 6,000-year history as a winemaking country attests to this. His precious white grapes, Obeideh and Merwah, took millennia to become accustomed to the heat, sun and drought of their landscape, where they ripen in October, two months after the other white grapes, because they need time to reach their fullest expression.

He found the best from every Musar grape if it stayed in the vineyard as long as possible in order to become the best expression of itself. The same in the winery: the best flavours would be extracted from the grape must if it underwent a long, slow maceration and fermentation. Serge would explain that in some years, his wines could take as many as five months to finish fermenting. The slower they were made, the better they aged, but the longer it took for them to be ready to drink.

Serge believed in taking time with every aspect of his wine out of respect to its heritage and its own innate character. Ageing in oak barrel, then in bottles in the cellar, takes place for as long as possible – until the wine is seven years old – before it is released for sale, so that no wine is drunk before it is ready.

And the 'Philosophy of Slowness' extends to tasting the wine too. Not a sip should be taken until the aromas are fully observed and understood. Then one should take time to notice and to feel the flavours – to let the wine unravel in the glass. 'Taste it now, then again a few hours later,' he would say. (He famously tasted a bottle of 1959 Musar Blanc, decanted and left at room temperature, over the course of a week. At first when he tasted it, it was undrinkable: it smelled sweet but tasted dry and austere. But the next day, with foie gras, it was 'amazing'. Then, over the week, time and oxygen played their tunes on the liquid and it evolved, oxidized and expanded. After seven days, the wine had turned from light tan to bronze, copper, then deep gold in colour, and

---

*Almost as cryptic as Serge's own elliptical pronouncements, this literally translates as: 'Time is as a sword. If you don't use the sword properly, you may wound yourself with it.' Or: 'If you don't spend time wisely, it will spend you.'

had gained the flavours of grilled figs, marmalade and candied almonds, becoming more delicious than ever.)

As Marc explains it: 'Musar wines give you the opposite of "instant gratification"; the opposite of any hectic metropolis way of life. You cannot expect to enjoy drinking them immediately. Our wines start off introvert, low and discrete. As they open up, they evolve and grow upwards. The more you wait, the more you get out of them. It is a journey, a dynamic experience, not a static taste. My father would say: "Give my wines more time and they will give you more pleasure." I like to drink them between 15 and 35 years of age, so they keep a little more punch. But Serge used to prefer them even older for the complexity they developed.

'He once told a young journalist to open a bottle of Chateau Musar white, put it next to her bed, and drink a sip every night before going to sleep over the course of one month. She tried to take notes on the wine, starting well in the first week, filling up pages of her notebook. In the second week she got confused as the wine was changing all the time; its flavours evolving and transforming, so she had trouble describing it. In the third week, she stopped taking notes and just enjoyed the wine.

'Our wines force the drinker to wait, to enjoy the moment, but also to be patient. If you rush, you miss out. This is a good rule for life too. We are often rushing to get to the next thing in our life and in doing so, we do not appreciate the time and space we are in now.'

Marc also tells of a time in the late 1970s, when the head of the Appellation Contrôlée of Burgundy came to Ghazir to taste the wines. He was with another Frenchman. During the tasting, they started counting: '12, 13', then '14, 15', then '16, 17'. Serge asked: 'What are you talking about?' The answer was Caudalies. In Burgundy, the length or staying power of the wine in the mouth is counted in caudalies (seconds). An appellation is granted when a wine achieves a caudalie from 10 or 12, but only rarely do wines last beyond 15. The men were astonished by the length of Musar on the palate. 'If wine does not linger, forget it.' Another Serge-ism.

What Musar does differently...

The reward of this 'slowness' was (and is) a wine that ages almost indefinitely in bottle, and that has a reputation for being at its very finest from 30 years after the vintage. Even in vintages as desperate as 1984, when war in Lebanon made it almost impossible to bring the harvested grapes back to the winery, the wine was able to come back to life 'like a phoenix' after a long period of quiet maturation.

Technically, it is hard to explain what allows these wines to become quite so expressive. But perhaps it is because they are cosseted but not

Gaston inspects the rows of neatly stacked bottles – each vintage is stored for five years this way before release.

Chateau Musar's beautiful old winery building was once a convent, with far-reaching views out to sea.

controlled in the same way (for example) as a Bordeaux, burgundy or Napa Cabernet would be? Serge and his son Gaston would say that the philosophy of 'minimal intervention', allowing oxidative conditions, minimal sulphur additions, with no fining or filtration, is the answer.

'We don't fine or filter our wines. This would be like removing a living part of them – they would be lobotomized. And who wants to drink a brain-dead wine? If you have the luck to work with something that is alive, you should never kill it,' said Serge.

Allowing the wine to respire is another factor. Inevitably, wines come into contact with air when fermenting, when being racked from one vat to another, when being decanted into barrels, when resting in oak and during the bottling process. Wine that spends a long time in bottle will breathe tiny amounts of air through its cork. An oxidized wine may smell odd, taste cooked, or even burnt, but a small amount of oxidation can be a good thing. It can lift and enhance a wine's aromas; it can give the wine longevity and resilience. In the case of Chateau Musar – whether red or white – it means that the wine gains a tremendous ability to withstand oxidization once the bottle has been opened and decanted, and not only this, it will then enhance and improve over time – a week, a month, even (Serge tried it) a year.

That oxidation is a factor in making Musar is not just a matter of choice. The three to four hour journey taken by the grapes from vineyard to winery – by the lorry-load from Beka'a to Ghazir – has the inevitable affect of introducing oxygen and minor bacterial reactions at an early stage. Refrigerated trucks have never been an option for the Hochars – nor are they now. And this journey is as much a feature of their wines as Cinsault grapes and Nevers oak. The end result can be a higher than usual note of volatile acidity (VA) in the bottle.

Most wines have VA, usually at undetectably low levels. In some it is seen as a fault (if too strong, it can show itself as a whiff of acetone or nail polish remover), in others, notably Australian Shiraz, port or Amarone della Valpolicella, it can be seen as an integral part of the wine, an asset, bringing balance and harmony. Serge described it as his secret weapon: 'VA helps a wine to be smelled. It is the volatile elements in any substance that we perceive as aroma,' said Serge: 'But it must be in harmony with the wine's tannins, alcohol and grape varieties. Even the place in which you are making the wine will affect the balance of VA it requires. But when it is right, it will add to the complexity of the wines and their ability to age well.'

Chateau Musar has the fruit concentration, body, alcohol and tannin to be able to balance VA and use it to the wine's advantage; it is part of its signature character, its complexity.

As Marc Hochar explains: 'VA may polarize opinion when observed in younger wines but the longer a wine ages, the more integrated it becomes. It settles with the rest of the characters, the acidity, the tannins and structure and becomes an advantage, it gives the wine life. It probably explains why Musar's aromas have the ability to jump out of the glass.'

Marc gives the example of the 1995 vintage: 'One of our vats had not finished its malolactic fermentation and the vintage was bottled at three and a half years; six months later than usual. When it was due to be released at year seven, the VA was so high the wine was unbalanced and difficult to enjoy. As

In the deepest part of the cellars, with cliffs for walls, lie the locked shelves of precious library vintages.

the 1996 was a lighter bodied and more accessible vintage, we decided to release both vintages at the same time. A year later, the 1995 vintage settled down and it became one of our most popular wines. Even today, when I put the 1995 in a blind tasting with other vintages it often comes out number one. It has a lot more VA than our other vintages, but this is integrated and gives the wine its presence. There are certain wine critics – one in particular who won't tolerant any VA – who always give it a low rating. And yet it is invariably the first wine to sell out when we release library bottles to the market.'

Given time in the glass, a Musar wine will change, evolve, and show you different, often unexpected facets of its character. It seems to adapt and harmonize with the circumstances it finds itself in. It is what Serge wanted it to be: an intelligent liquid that shares a conversation with you. And as with all the best conversations, this one does not remain fixed on one topic, but moves between subjects and ideas, answering questions posed by your palate and your memory, and equally, asking you how you feel about this smell, or that flavour nuance, or the way the light catches that part of the glass. Serge would say that the confidence to hold these conversations is found only in those wines that have the absolute trust of their winemaker. And the content of these conversations is a reflection of the winemaker's skill and ability to understand the difference between knowledge, which can be learnt, and knowhow, which comes through intuition and time.

To better understand this, we need to step backwards... ❧

Chapter 5

OLD AS TIME

The story of an ancient land

Chapter 5

SERGE-ISM

To taste an old wine, you have to give it time for it to tell you its life story,
because my wines interpret time, beautifully. I have learnt from them that
slowness is important. Slowness gives you the time to understand and
appreciate complexity. In wines and in marketing, simplification is the trend
but I prefer the complex over the simple. This is not only due to my age – we
are all ageing by the second – but I believe that ageing equals complexity.

SERGE HOCHAR

Serge Hochar, like many of his fellow Lebanese, was particularly proud of his Phoenician heritage. He enjoyed explaining that his homeland, Lebanon, was present at the birth of human history. It was here that the Phoenicians established some of the earliest organized cities – Byblos, Sidon and Tyre. They developed trade routes that spanned the Mediterranean spreading as far as Cádiz, Portugal and Africa; they invented the world's first written language as an aid to trade, and they grew the 'fragrant wines of Lebanon' mentioned in the Bible, exporting these – plus the sturdy, scented cedar wood of their forests – far and wide, to countries and cultures that grew to admire the pioneering, entrepreneurial spirit of these benevolent traders, who shared not just their produce but their knowledge too.

Serge relished telling the story of the Phoenicians and explaining that their influ-ence is still vibrant in Lebanon today. He believed that the Phoenician understanding of wine – rooted in his country and present in his blood – is what drove him forwards to continue their legacy.

The Phoenicians as traders...

The Phoenicians were an enterprising people. A Semitic race living in city states along the fertile coast of today's Lebanon and northern Syria, they would probably not have seen themselves as 'Phoenicians', but as Canaanites. (Canaan and Kanaan remain common Lebanese surnames.) They were at the height of their powers between around 1550BC and 330BC, when their combination of local materials, technology, curiosity and driving entrepreneurship made them the force that coloured the cultures of the Mediterranean coast.

'Phoenician' comes to us from the Greek *phoinos*, meaning 'purple', because the Phoenicians' first export business involved the murex, a large predatory sea snail that when

The Phoenicians were sea-going explorers, traders and entrepreneurs. Lebanese today are proud to tell their story.

boiled yielded the rich, luxurious purple dye sometimes called Tyrian Purple, after the city-state of Tyre. Murex-coloured cloth was the badge of the Hebrew, Athenian and, later, Roman aristocracy, and he who supplied the dye got rich too. The Kohen Gadol, high priest of the Hebrews, officiated in a prayer shawl dipped in Tyrian dye, a style echoed in the tallit still worn by Jews to pray. And, as Serge would say: 'As anyone who's ever seen the movies *Ben Hur* or *Gladiator* will have gleaned, Roman emperors were particularly partial to wearing murex-purple cloaks too.'

Of course, to export royal purple dye, wine or any other valuable goods, you needed first to master the sea. This, the Phoenicians did admirably. They built the world's first man-made harbour at Byblos, and developed a new style of ship that enabled them to open markets all around the Red Sea and the Mediterranean. Their 'bireme' was the jumbo jet of its day, a double-decker super-fast galley about 30 metres by seven, with a mast and a square sail, dedicated cargo compartments, and seating for 120 slave rowers. Their next-generation triple-decker trireme confirmed their domination of the Mediterranean.

As any international trader knows, conducting successful business requires more ingredients than buyer, seller, supply and demand; a clear, concise, unambiguous common language must be employed. So to add to their achievements the Phoenicians composed the world's first written alphabet. This, in time, became the basis of the Greek, then the Etruscan, then Roman alphabets. It's tempting to imagine that the world's first formalized writing was a poem of love, or a commandment from above; more likely it was a delivery note informing an Egyptian that he had 30 days in which to pay for his wine.

The first commercial winemakers

Wine is thought to have first been made over 8,000 years ago in the Caucasus Mountains

The Phoenician Empire at its fullest extent (below) and a revolutionary 'bireme' galley – the jumbo jet of its day.

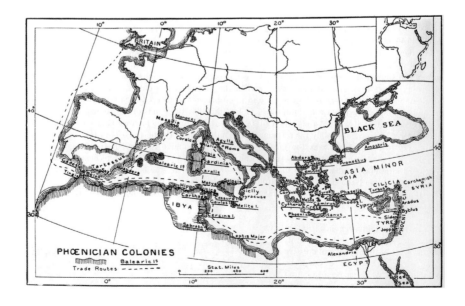

between the Black and Caspian Seas. Knowledge of this precious fermented grape juice would have passed to the Phoenicians via the trade routes of Mesopotamia that fed west to the Mediterranean coast – and with that knowledge went vine cuttings to get things started.

The Phoenicians were excellent farmers who understood exactly which sunny, south-facing vineyard slopes would produce the healthiest vines for winemaking. They took matters seriously, and the grapes they culti-vated were the world's first to be profession-ally grown and made into wine. According to ampelographers their cultivar *Vitis vinifera pontica* is very likely the parent of all native wine grapes grown today in North Africa, Greece, Italy, Sicily, Sardinia, Spain, the Balearic Islands and Portugal. Archaeological evidence of grape seeds and pollen traces plot the Phoenician expansion to all these areas.

Not only was wine popular for general con-sumption, it quickly became appreciated as an offering to gods, kings and deities. The Phoenicians, with their ever-expanding mari-time trade networks, designed and fired spe-cial terracotta urns in which to transport it to their markets. These narrow-necked 'Canaanite

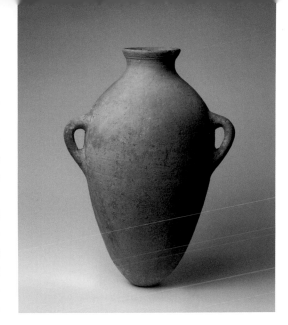

A 'Canaanite jar' for storing and trading wine.

jars' travelled all around the Mediterranean and beyond long before the Greeks created their amphorae. Examples have been unearthed in ancient Phoenician settlements and discov-ered in the debris around shipwrecks, some even bearing the name of the producer.

As trade boomed, the Phoenicians reaped the benefits. They created new markets along the Mediterranean coast, in Egypt and as far away as North Africa. They also began to colo-nize areas inland as they ventured down major rivers such as the Nile, Tagus and Guadalquivir.

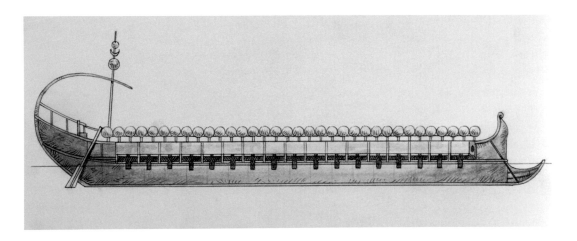

Egypt was an early, important export market for both wine and Phoenician cedar wood, and these intrepid seafarers soon understood that their wines, even when confined in a freshly-baked Canaanite jar, could spoil and oxidize during any sea voyage. Thus, they became practiced at adding a layer of oil to the top of their jars, to prevent air getting to the wine, and would sometimes flavour their poorer wines with pine resin, as Greeks still do with their retsina.

Connections with Egypt opened up trade links to Somalia for frankincense and Ethiopia for myrrh. Both these were essential ingredients in votive temple offerings, and passed through the Phoenicians' hands along with slaves, elephant ivory, and lion and leopard skins from the heart of Africa. Glass was first blown here ('all the better for drinking wine from', said Serge) and silver came from their

A springboard for the Mediterranean wine trade, Tyre was later besieged and captured by Alexander the Great.

colonies in Iberia. Tin came to them from Cornwall in Britain via Breton intermediaries. Phoenicians kept secret their recipe for the alloy of tin and copper that made their bronze more durable than the arsenic copper bronze made by their competitors in Mesopotamia.

Serge would pause at this point in his Phoenician story: 'May I tell you something confidential and controversial?' he would say conspiratorially: 'The name of Britain, as in Her Britannic Majesty, was given by our Phoenicians. In their language, Bar Tanac meant, Land of Tin.'

Masonry was another exportable skill, and the Phoenicians were credited with an almost occult knowledge of the art and science of building. King Hiram the First of Tyre sent timber and men to King Solomon for the building of his temple in Jerusalem. The Bible credits Hiram Abiff as the master-craftsman of the project (1 Kings 7:13–14 and 2 Chronicles 2:13–14) presented to Solomon as: 'A skilful man, endowed with understanding...skilled to work in gold and silver, bronze and iron, stone and wood, purple and blue, fine linen and crimson, and to make any engraving and to accomplish any plan which may be given to him with your skilful men and with the skilful men of my Lord David your father.'

To Carthage and beyond

Ten generations later, King Pummay of Tyre had his sister's husband (who was also her uncle) assassinated. She was Queen Dido, who, as told in the legends, quickly fled to what we'd now call Tunisia, where she founded the great city of Qart Hadasht – Carthage. Carthage became the Phoenicians' most successful colony, a city that ruled much of the western Mediterranean for nearly 500 years, until it became so great a threat to Rome that the Romans felt compelled to destroy it in the last Punic War of 146BC. Carthage was described as having countrysides full of grapevines and olives. Its greatest writer, Mago, is known to have composed manuals on farming and viticulture showing methods of planting similar to those used today.

Just as Qart Hadasht became 'Carthage', the Phoenician colony Gadir in Spain became Cádiz. Similarly, the names of Malaka (Málaga), Tingis (Tangier), Karalis (Cagliari, Sardinia) Eryx (Erice, Sicily), Phoenice (Venice), Malat (Malta) and Setúbal (Portugal) all retain echoes of their Phoenician past, as does the great continent of Europe, whose name comes from the Phoenician 'Erob', meaning where the sun sets.

The Phoenicians are known to have taken wine as far as Portugal to trade it with local silver and tin. Their winemaking equipment has been discovered in Valdepeñas in Spain, where they ventured inland along the River Ebro. And in founding Cádiz (Spain) as early as 1100BC, their wines – not Roman wines – are likely to have been the forerunners to sherry (a probability that raised a triumphant bushy eyebrow from Serge).

More than a mere beverage

Serge would continue his lecture saying that the Phoenician legacy was about much more than the spread of viticulture, it was about spreading knowledge. They were a peaceful people who valued the development of relationships beneficial to all. They admired the traditions, culture and religions of all they encountered, and would always endeavour to treat other societies with respect. They were known to have taught their winemaking and ship-building skills directly to the Greeks and encouraged their expansion beyond the shores of the Aegean. They were generous, they were diplomatic and they were forward thinking.

Professor Pat McGovern is a Biomolecular Archeologist at the University of Pennsylvania Museum. His speciality is fermented beverages and his research enables us to know more about wine in everyday Phoenician life. He says: 'The wines of ancient Lebanon and Syria were much more than a technological marvel. They were the social lubricants to break down barriers, stimulate music and the arts and bring people together. They were the supreme religious symbol for prosperity and well-being, in life as well as death. And they were medicines for dissolving and dispensing botanicals, to rest the weary, relieve pain and cure disease.'

Serge agreed with this. He said: 'My wine is my doctor and my best friend. When you are happy, it makes you grateful. When you are sad, it makes you know that sadness can end. When you are in love, it makes you intoxicated with love. When you are not in love, it gives you the capacity to love.'

Farewell to Phoenicia, hello Greece

By 333BC the Phoenicians were a waning force in the Mediterranean and Alexander The Great was on the hunt for new Greek territories. His empire occupied most of the land from the Aegean to Afghanistan, and he had recently subjugated the Turkic tribes and Persians in Asia Minor – where Darius, King of Persia, had fled the field of battle, leaving his family to enjoy some painful lessons in Macedonian hospitality. As Alexander's army swept inevitably into Phoenicia, it met little resistance from the locals, that is (in Serge's words), until he tried to cosy up to them.

Here in the story, Serge would add firmly: 'We are a contrary people. We are proud and headstrong, yet somehow we are also soft and patient. A Lebanese can be your best friend and your worst enemy very quickly. We love to argue and we love to be hospitable, too. But if there is one thing we Lebanese distrust, it is an invader being nice to us.'

Soon after his invasion Alexander, in celebratory mood, strode into Tyre's temple and attempted to make an offering to Eshmun, the city's god. The Tyrians were enraged by this act of conciliation. Alexander reacted in the way he knew best, by withdrawing, then laying siege to the city. Tyre was built on an island and was well fortified, but when it came to war Alexander wasn't called The Great for nothing. After seven bloody months under siege Tyre capitulated, Alexander having built a causeway to the fortifications before breaching them. Those of its inhabitants who didn't die during the siege, or in the massacre immediately after it, were sold into slavery. It is believed that 30,000 men, women and children became Greek slaves overnight. This may seem like a tough, if decided, way for Alexander to gain a city's trust. But the other Phoenician cities saw immediately that he was a fine fellow and requested that they become Greek citizens forthwith.

Alexander died young, at 32. His empire was divided up between his generals and their progeny. Ptolemy the Saviour got Southern Syria and Egypt; One Eye Antigonus got Europe, and Seleucus the Victor ruled over Phoenicia, Asia Minor, Northern Syria and Mesopotamia. Predictably, the generals bickered and fought between themselves, with Seleucus busying himself with kicking Ptolemy's backside, and Antigonus driving Seleucus out of Babylon. Seleucus' son, Antiochus the Saviour, having married his stepmother, steadied things down for a while, and the Seleucids reigned over Phoenicia more or less peacefully for the next 250 years. During their reign, Phoenicia was renowned as a place where poets, philosophers, winemakers and wine drinkers thrived.

The arrival of Bacchus

In 64BC, the Roman emperor Pompey decided to annex Seleucid Syria and Lebanon to his empire. Then followed a period of quiet prosperity as the people of Byblos, Tyre and Sidon were granted Roman citizenship. Trade relations improved, and new wealth accrued from

The atmospheric Greco-Roman ruins at Ba'albek – among the finest remaining temples anywhere in the world.

THE WONDERS OF BA'ALBEK

The Temple of Bacchus, the largest known Roman monument to wine. Despite the ravages of history, earthquakes, vandalism and theft, it remains almost complete, with detailed carvings of grapes still visible. The Hochars take much inspiration from this site.

exports of cedar, pottery, jewellery, perfume, wine and purple dye to Rome, paved the way to a period of widespread construction and urban development.

The Romans admired the strong legal foundations that had been developed by Phoenician society, and accordingly Beirut Law (known as the 'Mother of Law') became one of the three schools of their own Roman legal system.

They also admired their places of worship, and as usual their brutal tendencies towards territorial conquests became noticeably softer when it came to temples. In Lebanon's Beka'a Valley, the millennia-old temple complex at Ba'albek was not only treated with sensitivity, it was enhanced beyond that of any other Roman site. Where the Phoenician god Ba'al had once been worshipped became Heliopolis, 'the City of the Sun' under the Greeks. Then, under the Romans, it was reconsecrated to their gods Jupiter, Venus and Bacchus.

Ba'albek remains today the largest temple in the Empire. A site of 700 hectares, it comprises

The Romans admired Ba'albek as a place of worship, and embellished it with their own characterisitc structures.

a hexagonal forecourt, the great court featuring a square alter, the Temple of Jupiter (measuring 48 by 88 metres) and the still largely intact Temple of Bacchus (66 by 35 metres, 31 metres high, with eight 20-metre columns along each end, 15 along each side). The temple of Jupiter, once surrounded by 58 such columns (six remain standing today), would have been considerably larger than the Parthenon in Athens which measures 69 by 31 metres.

As a feat of engineering, Ba'albek continues to baffle the finest scientists. No modern engineer has managed to work out how the 800-ton quarried stones that provide the base for the temple could have been moved there, by the Romans or (as must have been the case) by the cultures that preceded them – let alone withstand 2,000 years of earthquake and neglect.

And why such magnificent temples should be located here exactly, nobody is certain.

Although this does mark the confluence of two major trade routes, and an important water source, there is no obvious reason for its high standing as a religious site. It may be that the orientation of the sun, the stars and the seasons can provide a reason. The sheer impossibility of moving the giant stones even suggests the intervention of some kind of unworldly force – or, according to one legend, 'giants'.

One thing is certain, people have worshipped here ever since there were people. And the fact that one third of this ornately carved temple is devoted to Bacchus – the god of wine – is something the Hochar family take great reassurance from. Serge's answer? 'This is the only serious temple erected to Bacchus anywhere in the Roman world. And they put it here, in the Beka'a. Why? Because the Romans and Greeks, the Phoenicians and the Minoans, and all peoples who came before them, all knew that the Beka'a is the spiritual home of wine.'

His sons, Gaston and Marc, also draw a certain strength from this place. 'I was not sure about making the family business my life; about becoming part of the Lebanese wine tradition,' says Marc: 'But when I came to Ba'albek, it all suddenly made sense. This place is at the heart of our wine culture – it shows our history with wine goes beyond any other – and it is important to represent it with the unique wines we produce at Chateau Musar.'

The Phoenicians live on

The land we call Lebanon today eventually became part of the Eastern Roman Empire, ruled from Constantinople and known as the Byzantine Empire. The years of Byzantine rule saw the gradual retreat of the god Bacchus as Christianity took hold, but wine remained part of everyday culture.

The Byzantine era came to an end in 636AD with the Arab Conquest. Thereafter, Muslim and Christian communities began to live side by side in the occasionally uneasy relationship that still continues. Under Arab rule, alcoholic drinks were prohibited but production never completely died out. They, and wine, were given a boost by the arrival of the Crusaders – who found the wines of the Holyland 'very good, and very noble' – and thrived again during the Ottoman Empire. In his book *Mount Lebanon: A Ten years' Residence from 1842 to 1852, volume 1* (1853) Colonel Charles Henry Churchill describes the wines of Lebanon as 'hardly surpassed for richness of colour and delicacy of flavour' with 'a perfection not attainable even in the South of France'.

The boundaries of Lebanon today remain very similar to those of the old Phoenicia. And though their era has passed, Phoenician DNA lives on in many Lebanese people. Serge Hochar would now call in the example of Tarek Sakr his winemaker and 'spiritual son'. This talented, charismatic man who now makes the family's wines, tested his DNA and found that he was authentically Phoenician. He certainly looks the part: just like the figurines of warriors in Beirut's National Museum.

The dynamic Phoenician spirit is still very much to the fore in Lebanon too. Whether animated by promising business opportunities, forced by economic necessity or driven out by war, Lebanese people still travel widely abroad. The Lebanese diaspora is approximately 20 million people, of whom around 10 million live in Brazil alone. The population of Lebanon today is less than five million.

Even in Phoenician times Lebanon was an ethnically and religiously mixed society of Canaanites mixed with Yemenites, Jews and

Philistines. In very few countries is there such a broad cultural mix, with so many languages spoken. Today, there exist 18 officially recognized religious communities, all of whom have a greater or lesser representation in national political life.

And the Phoenix rises

Despite appearances, there is no etymological link between the words Phoenix and Phoenician. But circumstances would somehow prove that there is an association: the Phoenicians were renowned for their recovery from disaster. When invaders plundered their coastal cities, raised them, robbed them, they patiently returned and rebuilt them...

In the Egyptian Book of the Dead the phoenix is represented as a stork or heron-like bird that reminded its people to honour the morning sun, and whose story and purpose was linked to the sun god, Ra.

The Greeks, excellent storytellers that they were, told a richer tale. The phoenix was a purplish coloured bird, the size of a hawk. It fed upon frankincense and myrrh and lived for between 500 and 1,000 years. Towards the end of its life cycle, it would build a nest of fragrant twigs, cinnamon and thyme, frankincense and myrrh, which would at some point self-combust. From the ashes of this fire would rise another phoenix.

In the fifth century BC, the Greek historian Herodotus explained that he'd never seen a phoenix, but had seen a painting and heard a myth retold that he didn't quite believe. The myth ended with the new phoenix embalming its father's ashes in myrrh and flying them to Heliopolis in Egypt for burial. Other Greek writers accepted this tale wholesale, as did the Roman poet, Ovid. They also placed the phoe-

nix firmly by a well in Phoenicia. Each morning it would bathe there and sing a hymn to the new day. The sun-god Helios tarried each morning to listen to the bird's song, so beautiful was it.

This curious mythological creature was remembered during the Israeli air attacks of 2006, when bunkered deep below ground in

Balanced view: Marc Hochar admires the temple ruins at Ba'albek that inspired his belief in Lebanese wine.

the Chateau Musar cellars along with his workers, Serge opened a bottle of 1974 red. This he poured into glasses, allowing (as ever) that it should be given time to breathe, to gently get to know its new surroundings, it having spent 30 years inside its thick green glass bottle. His voice would have risen as he proposed a defiant toast to his battered country: 'Lebanon is like the Phoenix... You can bomb us, invade us, burn our villages and fight your wars here. You can destroy us seven times, and each time, Lebanon, like the Phoenix, will rise.' 🏵

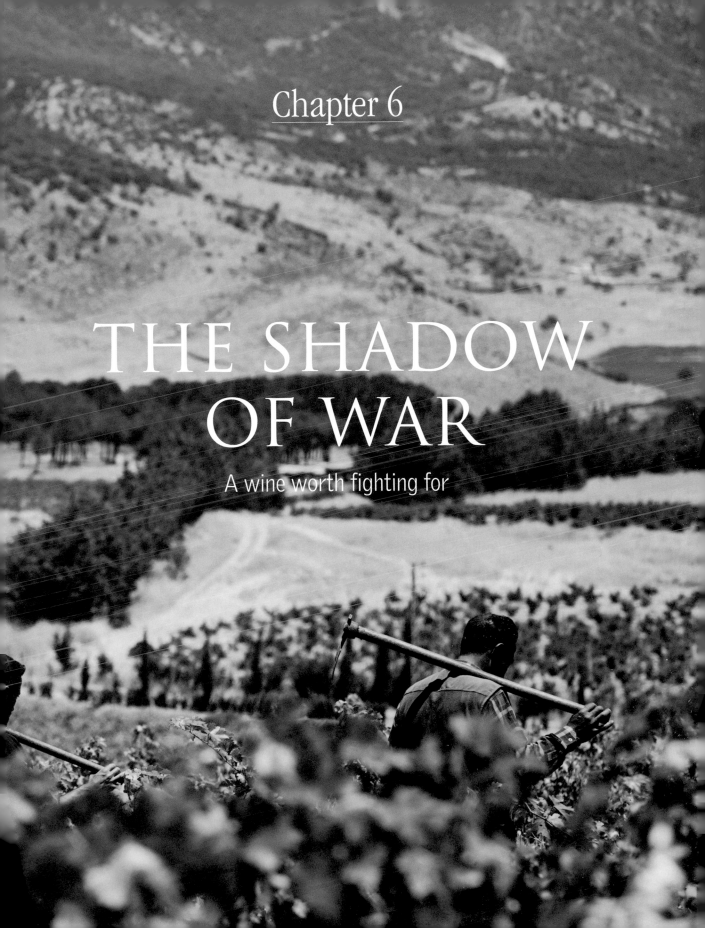

Chapter 6

THE SHADOW
OF WAR

A wine worth fighting for

Chapter 6

ANACHRONISM

A recipe for disaster: start with a generous serving of rich Maronites. Take one dispossessed Palestinian population. Seethe with resentment. Add to this a mixture of revolutionaries, Arab nationalists and anarchists. Stir. Whisk a non-functioning, collapsing government together with international ignorance to achieve an insubstantial, collapsing froth. Season with poor unemployed youths of every Muslim hue and creed. Boil together with fresh Druze from the Chouf Mountains. Garnish with Syrian, Iranian, Iraqi and Israeli know-how, money and weapons. Keep heating until the taste is bitter and sour, and the dish explodes. (This it did on April 13th 1975.)

KEVIN GOULD

THE SEEDS OF LEBANON'S Civil War were sown after the Great Famine of 1915–18, which led to the starvation of over 200,000 Lebanese on Mount Lebanon. At the end of World War I and at the collapse of the ruling Ottoman Empire, the French took up the Lebanese mandate and decided to move the Lebanon-Syria border to the east of the Beka'a Valley, positioning it at the foot of the Anti-Lebanon Mountains. This would not only maximize the area under its control but gain Lebanon a defensible border and enough agricultural land to feed its people.

The trouble was that this changed the demography of the country's population completely. The peoples of the newly annexed territories were predominantly Muslim or Druze. Lebanese Christians (mainly Maronites, but also Orthodox and Catholic) now made up

barely 50% of the population. When it came to drawing up Lebanon's constitution in 1926, a confessional system was developed ensuring that government seats and civil service positions were allocated among religious sects according to their estimated population ratio. This was supplemented at Lebanon's independence in 1943 by a 'gentleman's agreement' that its president would always be a Maronite Christian, it's prime minister a Sunni Muslim, the speaker of parliament a Shi'a Muslim, and the deputy speaker a Greek Orthodox. But the French constitution gave the president the power of veto over any legislation, and therefore guaranteed Christian dominance in the event of any future change in the population distribution.

As events unfolded, the Muslim population was to increase dramatically. But no census

Sunni Muslim

Shi'a Muslim

Maronite Christian

Greek Orthodox

Druze

Greek Catholic

other Christian

The major religious groups of Lebanon (2012): political stability has long been affected by sectarianism here.

was allowed to confirm the new ratios. Mistrust of the country's political system grew.

Matters weren't helped in 1948 when Israel proclaimed Independence, and, while the US and the UN looked diplomatically elsewhere, began to buy and bully the Palestinians out of their villages and farms. Of 900,000 Israeli Palestinians, 80% or more left or fled. Many of these went north to Lebanon.

More fuel to the fire

This arid region depends not on oil, but on water for its survival. In the 1950s and 1960s, the River Jordan occupied centre stage in the arguments. Israel wanted to divert a third of it to irrigate the Negev Desert. Unsurprisingly, Egypt, Syria, Jordan and the Jordanian/ Palestinians on the West Bank didn't want them to. There were some meagre attempts at international conciliation, but Israel went ahead anyway, and by 1964 had completed a pan-Israeli water grid designed to green the desert – and (ultimately) provide Europe's new supermarkets with Jaffa oranges and avocado pears. In retaliation, Syria and Jordan started diverting River Jordan water at its source, close to where St John is said to have baptized Jesus Christ. In April 1967, the Israeli Air Force launched air raids to stop them. Two months later the world saw the Six Day War.

Caught unawares, all neighbouring states lost heavily to Israel. Egypt lost Gaza and the Sinai; Syria lost the Golan Heights; Jordan lost East Jerusalem and the West Bank. As many as 145,000 more Palestinians found themselves in Jordan. They were not welcomed by the Jordanian minority.

Pity the handsome King Hussein. Swaggering Palestinian fedayeen ran their own militarized fiefdoms throughout his kingdom. They didn't care that he was of the noble house of Hashemi whose line can be lawfully traced back to the Prophet himself. Palestinians were the law, now, and peace was no longer a language spoken in Jordan. The Palestinians launched guerilla attacks into Israel and at the Jordanian army; extorted ruthless 'taxes' from anyone they could; manned illegal checkpoints; openly carried arms, and openly challenged the ruling minority. Hospitality is a cornerstone of Islam, but by 1970 the Palestinian houseguests had worn out their welcome and had decided to become head of the household.

Sandhurst had taught Hussein many military strategies. But the Jordanian Palestinian problem was compounded by the fact that Hussein's army wasn't fighting a single enemy. The PLO was a sort of loose holding corporation. Corporate HQ was Amman, with regional offices in Beirut, Damascus and Cairo that never quite controlled a ragtag of loyal and less-loyal branch companies (some of which splintered into further factions). As was the chic in 1970s radical movements, these factions were identified by acronyms. As well as Fatah and As-Sa'iqa, Hussein had to contend with the DFLP, PFLP, PPP, PLF, PDU and PAF.

Things deteriorated. The Palestinians gathered support from the Iraqi army and from Syria, who mobilized a tank brigade, the PLA (Palestinian Liberation Army) based in Damascus. A series of airline hijackings involving British and American civilians brought in the US 6th fleet and two UK aircraft carriers to back up Jordan. Not to be outdone, the Russians, firm friends of Syria, muscled-in with 20 warships and six submarines just 50 kilometres up the coast.

On September 27th in Cairo, on the day the Hochars and their loyal Bedouin workers started gathering their grapes in the Beka'a, King Hussein and Yasser Arafat signed a conciliatory agreement allowing the Palestinians a degree of autonomy, but the very next day Egypt's President Nasser collapsed with a heart attack. With the PLO's political protector-in-chief dead, the Jordanians renounced the agreement and renewed their bid to oust the Palestinians from their country.

By November 1971 they had succeeded. The PLO was forced to regroup and decided to relocate its HQ in southern Lebanon, where they found allies in the local Shi'a population. Their presence brought the number of Palestinians in Lebanon to nearly half a million,

Faces of war: Druze militiamen pose for photojournalist Claude Salhani in Beirut, June 1976.

further changing the demographic ratio to the disadvantage of the Christians. It also prompted a fierce reaction from the Israelis, who held the Lebanese government responsible for Palestinian actions. But Lebanon was unable to control the seething anger of the PLO, and nor could it defend itself against the resentment of Israel.

As well as the growing tension between its neighbours, Lebanon had to contend with growing internal hostility developing between its two main political coalitions. The Maronite Christians, or 'Lebanese Forces', led by Bashir Gemayel, were in favour of continuing the confessional system. Opposing them was an amalgam of groups sympathizing with the Palestinians (collectively known as the

Lebanese National Movement) who demanded reforms that would benefit the Muslims.

In 1975, battles between Christian militias and the PLO spread to Beirut, resulting in a division (the 'Green Line') between the east and west parts of the city and causing a bitterness that was to dominate the region for the next 15 years, leaving Lebanon in ruins.

The fight for survival, 1975

Between 1975 and 1990, Lebanon was a mosaic of wars. Each time a clear picture began to emerge from the fractured pieces either a vital fragment was found to be missing, or another petulant force would smash the tesserae to smithereens. Soon enough, no-one could remember the original image that the mosaic was meant to resemble.

Fighters, foreign and home-grown, spread their anger everywhere, and this rapidly

Israeli tanks besiege Beirut, June 1982, in a concerted effort to neutralize PLO bases in and around the city.

infected the entire country with fear – and more anger. Your surname alone was often your death sentence. Your accent would betray your ethnic origin. Being in possession of either your name or your tongue at the wrong time or place could mean summary execution, protracted siege or bombardment.

'You talk of war, and of civil war,' Serge would say. 'There is nothing civil about war. What happened in Lebanon is that people fought their wars on our land.'

On a good day, the ride from the Chateau Musar office in Achrafiyeh at the centre of the city to the airport might take 15 minutes. But good days were rare in 1975. That August, there was peace in the Beka'a: war was confined to downtown Beirut. The Musar wine press had broken down just before the harvest and the replacement part was sitting over at

the airport. Without it there would be no wine. 'It was a beautiful day,' remembered Serge: 'The kind of summer day that grapes love.' (When Serge talked of the war, a sharpness entered his eyes. This was neither Hemingway machismo nor the look of an us-against-them, set-jaw partisan. It came from the memory of a daily life you simply had to endure.) 'En route, I noticed that outside the Sabra and Shatila refugee camps there was a militia checkpoint. We called these "*barrages de mort*", killing barricades. If you had a name like mine, they'd kill you – finished. I carried on to the airport, picked up the machine part and sat quietly by myself for 15 minutes. I thought, I know my faith. This is not a religious thing. But it is the same faith I have in my wines. So, I got back in the car. By the time I drove past Sabra and Chatila, the checkpoint had disappeared.'

Serge got back to the winery and told his employees that their jobs were safe, but that the war would last, and to get through it they must work seriously. They did.

By October 24th, the Hochar's 1975 harvest was safely gathered in. The grapes had been separated from their stems and the gash-red flesh had been turned into pulpy juice. This juice was now quietly fermenting in huge concrete tanks in the calm of the Ghazir winery. Back in Beirut, 25kms away, something altogether different was fermenting.

Beirut's smartest property has always been found on the Corniche, a strip of coastline between mountain and sea, where the elegant liked to stroll, and the wealthy reside in the smartest apartments. Here, like concrete fingers pointing skywards, a cluster of smart international hotels had grown tall. These towers, some as yet unfinished, became the ideal spots from which to throw bombs into the surrounding Christian area. The St George, the Phoenicia, the Holiday Inn, Palm Beach, Alcazar and Normandie hotels had all hell shot out of them in the seven weeks that followed. (The Holiday Inn remains today a bombed-out hulk whose empty windows stare sightlessly out to sea, standing as a stark reminder to all who see it.)

The Battle of the Hotels was the first crack in the schism that would separate Christian East from Muslim West Beirut for the next 15 years. The dividing line between the two, the Green Line, became a no-go area that came to symbolize the depravity of all urban wars. From the Chateau Musar offices in Beirut it is a 10-minute stroll to the Green Line.

TIMELINE

1976 30,000 Syrian troops invade Lebanon ostensibly to restore peace, but in reality this is Syria's attempt to claim the lands it believes it was owed when Lebanon became independent in 1943.

'In war, at every moment you can be in danger. War is not easy. And our wars were not short.' said Serge: 'In 1976 there was total war in Lebanon. We had no electricity. No fuel. No transport. No harvest. No nothing. 1976 was the only year in which we failed to make any wine. You cannot have a vintage every year. For others the weather is a problem – for us, it is war.'

Without the ability to reach the grapes, to pick them and bring them back to the winery, there could be no new wine. Even if there had been, there was nobody to buy it. So Serge saw the opportunity to reach out and find new markets for the wines he had in his cellars. His father-in-law

had a travel-agency in London and agreed to start importing his wines in the UK.

1977 From 1977 to 1990 (and during plenty of small dirty wars since) the Hochar family successfully made wine each year – often plucking grapes from the barbed wire zones of the frontline. 'During times of war, we have to put all our belief and all our assets in wine!' said Serge. 'We have to keep going with the things that are our essence. By now, I was used to war. So I kept on making wine; I was making it for a market I did not yet know existed.'

1978 The UN steps in with peace-keeping troops and Israel is forced to withdraw from Lebanon. Syria steps in instead, targeting the Christian population in the 'Hundred Days' War', the worst offensive for two years. Serge: 'There was very heavy shelling in Beirut. The Syrians shelled

Peaceful murals painted by Beirut students in 2009 replace the angry political slogans of the 1970s and '80s.

Achrafiyeh, where our office is. At the time, my wife Tania was running the Godiva chocolatier nearby. She took the kids to the winery cellar in Ghazir in shock. There were many dead in Achrafiyeh. My secretary hid in the strong metal filing cabinet. Many people here were psychologically affected.'

1979 Revolution in Iran leads to radicalization of the Shiite movement in Lebanon, and creation of the Amal party, the 'movement of the dispossessed'.

1980 Bashir Gemayal unites Lebanon's Christian military factions, creating the Lebanese Forces political party. Serge: 'This was one of the worst years of the war for us. I used to write a harvest report, but have mislaid it for this year.' Chateau Musar UK was officially established this year.

1981 Relations between Syria and Israel deteriorate. Serge: 'This was a year of terrible hardship. Tania and the kids left for London. They had to. I could not guarantee their safety. They came back,

then left again in 1983. I promised that I would join them the moment the war ended. We did see each other every few months when I could get a flight'.

1982 Israel invades Lebanon again with the aim of dislodging the PLO. Some 18,000 people, mostly civilians, lose their lives as the battle-zone moves towards Beirut. Italian, French and US troops assist in the evacuation of the Palestinians.

In the Beka'a, 80 hectares of Musar vineyards become the frontline between the Syrians and Israelis, whose tanks faced each other over the vines. Serge calculates that he will not be able to harvest grapes from them again until 1985. But he does. In the confusion that ensues after the invasion, the loyal Bedouin pickers collect what fruit they can and the Hochar trucks manage to make their way to the winery. Serge: 'The 1982 is a pure wine of war.'

1983 Times darken further as suicide bombing reprisals shock Beirut and its suburbs. In the Chouf mountains, the Mountain War begins. That summer, Serge took his family on holiday to the US. During a stop-over in Paris on their return journey, war broke out again so the family decided to settle in France.

Winter is severe in the Beka'a, with metres of snow coating the vineyards; summer is barely warmer and the Hochars' harvest was late. A break in fighting, instilled by the American fleet anchored off Beirut, came at just the right time for the vineyard manager to pick a few grape bunches and smuggle them into Beirut for Ronald to check. They were good: there had been no rain, no heatwave, and the grapes had reached perfect condition. Ronald put in a call to Serge, who was visiting the US, and the brothers decided to match the bravery of their vineyard manager and order the harvest in. 'It was very dangerous,' admits Ronald: 'Serge flew to Cyprus, then took a six-hour hovercraft crossing to Beirut; he arrived at the winery in Ghazir moments

after two rockets blasted the coast road he'd just been driving on.' The truck drivers, with their precious loads of hand-cut grapes, carefully negotiated the capillary-like country roads that would lead them to the winery, relying on a network of local gossip to learn which roads the militias were controlling with checkpoints and which route between the Beka'a and Ghazir would be the least bloody. They were successful. As Ronald said, there were two advantages: 'Our usual headache, the traffic jams, were gone. And the harvest, just like that of 1982, was excellent!'

1984 The Lebanese Forces, having controlled the capital since 1982, are expelled, and the Amal Party takes control of West Beirut. Peacekeeping forces (from the US, Italy and the UK) leave Lebanon.

Serge: 'We did our best with what we have, as always. This is the Lebanese way. You give me grapes and my job is only to help them to be the best wine they can be. In 1984 the sun was very hot and the fighting was hotter. After waiting and waiting for a break in the shelling, and more than a month after the last day of the harvest should have been, we quickly picked whatever grapes were left on the vines. Most were very ripe and sugary. There were enough grapes for only two truckloads, and only two of our drivers were brave enough to attempt the journey to Ghazir.

The first truck managed to find a way through the cedar forests in the northern Beka'a and get eventually to Tripoli, near the northern border with Syria. In five days it was with us. The second truck drove south over mountain tracks to Jezzine, then crossed the battle fronts on the way down to Sidon. From there, the trucks needed to avoid Beirut, so they waited for a boat. There was a terrible storm, which delayed the ferry to Jounieh. Eventually the storm was quiet enough and the ferry sailed slowly up the coast. The trucks arrived at the winery after seven days, on October 20th, 45 days after what should have been the end of the harvest – 45! So, in 1984 the grapes that we received were hot, bruised, sticky and very much fermenting. As an act of defiance, and as an act of faith, as a way of showing that the Lebanese spirit can never be broken, we made those grapes into wine. I made the 1984 to declare war against war.'

In 1984, Serge Hochar was named as the first ever *Decanter* magazine 'Man of the Year'.

1985 The Israelis continue their withdrawl from the south of Lebanon under armed pressure from Hezbollah. They keep an occupied 'security zone' along their border. In the War of the Camps, Palestinian refugee camps are targeted by Shi'ite and Amal militia. Serge: 'The Israelis continued to withdraw from Sidon but it was still difficult to get the grapes from the Beka'a to the winery. After this year, things started to settle down, on our terms.'

1986 Serge: 'Military action destroyed some of the vineyards on Mount Lebanon. Our white grape Obaideh was, and is, grown on the slopes of the Anti-Lebanon and was still available – but we only ever used this grape for Arak (Lebanon's aniseed-flavoured aperitif). Prior to 1986, the Chateau white was only produced from Merwah grapes. From 1986 we began to blend Obaideh with Merwah and the combination worked. We were the first to use these two local varieties in a unique new style of wine.' Close to four decades later, Musar is still the only one.

1987 Anger at the continued presence of Israel in Lebanon is augmented by the first Intifada (Palestinian uprising against Israel's occupation of the West Bank and Gaza).

1988 Following Amin Gemayel, Lebanon's new interim president is announced as Michel Aoun.

Serge: 'Life in Lebanon was not so hectic in 1988. It was a normal year, politically speaking.'

1989 President Aoun declares war on Syria. After seven months of fighting, with 800 dead, the Arab League negotiates a ceasefire. The Taif Agreement is signed, with the ethos 'no victor and no vanquished', in an attempt to end the Lebanese Civil War. Life is beginning to return to normal in the Hochar vineyards. In an impressive vintage, Carignan grapes begin to win Serge's heart.

1990 The terms of the Taif agreement are legalized, with reforms including a larger parliamentary assembly, an even Christian to Muslim ratio and reduced power for the presidency.

The final Syrian offensive on October 13th forces President Aoun into exile, and a new unified government under President Elias Hrawi begins the delicate job of piecing Lebanon back together.

Serge: 'We were told that the war had ended, but my nose said there were problems. We usually started to harvest our red grapes at around September 15th, but I was afraid of the situation as the Syrians were threatening General Aoun in Lebanon. So, we started harvesting on September 5th. Lucky we did, as 20 days later the Syrians attacked and blocked all the roads. We had finished harvesting the day before.'

In 1990, with peace in the air, the EU asked for evidence that Lebanon was a 'wine producing country' in order for Chateau Musar to be able to officially export. Lebanon had produced wine for 6,000 years. This EU directive required a law to be passed in Lebanon giving winemaking the status of an officially sanctioned business – a sensitive subject given that the Minister of Agriculture was a Hezbollah member of the Shi'a community. But the law was passed, and Serge became Lebanon's delegate to the OIV (Office International de la Vigne et du Vin).

The phoenix rises

After more than 15 years, Lebanon's Civil War was over. The foundations for its new republic had been laid. But the costs were high: more than 100,000 people had died, nearly one million displaced, and billions of dollars worth of property and infrastructure had been destroyed.

Musar's vineyards had suffered damage from fighting in the Beka'a, and the winery in Ghazir had been subject to intermittent shelling, but thankfully escaped major damage. As Serge said: 'If you ask me on which side I fought during these 15 years, I can only tell you that I was fighting for my wines. The wars were futile, useless. Without point. They were not only about religion, or power, they were about money. Militias at checkpoints demanded heavy bribes. And as soon as the fighting started in Beirut, different factions broke into the banks, blew up the safes and ran away with millions of dollars. Most militias imposed "taxes" on the people who lived in or travelled through the areas they controlled, and Musar's

Israeli soldiers disrupt the ever-fragile peace by shelling Hezbollah positions in southern Lebanon, July 2006.

trucks going through checkpoints were no exception. War is about profit.' And loss.

The sheer effort required to get the Musar grapes to the winery each year had been extraordinary. The pickers, who would often pick under artillery and gunfire as the dug-in militias bombarded each other, emerged as heroes. As did the truck drivers who risked their lives during every minute of a journey that in peacetime can be made in two and a half hours, but during war had taken five days.

All wineries like their grapes to arrive at the winery in the shortest possible time after they have been picked. This is because if grapes are bruised, or crushed under the weight of other bunches, the sun warms them up nicely as they travel and they start to ferment. A delivery of warm, fermenting grapes makes a musty, unattractive wine. So winemakers will try to prevent this happening by picking in the

Hope for the future? The winery at Ghazir, where modern machinery is put to more peaceful purposes.

cool air of night, under floodlights. Others load their freshly picked grapes straight into refrigerated trucks. Neither of these sophistications were available to the Hochars during war time.

As Serge had said: 'There was almost no vintage in 1984.' Just two trucks made their way back to the winery – one taking five days. 'My 1984 red was made from grapes that were so over-ripe and had travelled such long journeys to the winery that they were already hot and fermenting. The wine I made from them was like no other wine I have made before or since. I bottled it for the hell of it, and I trusted the 1984 to become whatever it decided to be.'

In wine terms and in human terms, Chateau Musar's 1984 red is the equivalent of sticking two fat fingers up at the twisted egos, tortured logic and misplaced hubris that cause men to fight and kill. By one of those vinous miracles that allows Musar wines to live long, healthy, interesting lives, a wine that should not really have been wine at all became not only a drinkable wine, but a delicious, complex exciting one, brimming with life and hope.

'I used to call it my Madeira, then it became like a port. But by 2013 what had been an oddity, an aberration, had become a fully fledged Chateau Musar, a beautiful wine, an elegant wine, a wine of the brain and a wine of the mind. We launched the tiny amount that there is of this wine on the market in 2014, exactly 30 years after it was made.' (It starts sweet-ish and strong, initially more like a Madeira than a classic Musar, but then arrives the recognizably Musar nose of warm, spicy dried fruit. Given time in the glass, it reveals a spine that is serious, persistent and dry.) 🐝

JOURNEY OF TASTE

Memories of Serge Hochar by Elizabeth Gilbert

S E R G E - I S M

When you taste my wine, you are tasting millions of things about it. You are tasting everything that has happened to that wine during its life until you drink it. How the weather acted upon it day by day when it was still grapes. How it fermented itself. How life was for it in the cask. What was happening in the world outside it when it was bottled. The way it reacted to travel, and changes of temperature. The way it is reacting and adapting to the place you open it and the glass you pour it into. The way it is adapting to you and your body and your brain and your mood. When I talk of millions of things, I mean, billions.'

SERGE HOCHAR

ELIZABETH GILBERT met Serge Hochar in 2004 and found a self-described 'ordinary wild person', a philosopher, an existentialist, a man in touch with both his own senses and those of his extraordinary wines...

One morning in 1990, a Lebanese winemaker was doing some paperwork in his Beirut apartment when shells started falling from the sky and the buildings around him began to shake. It was another Syrian assault.

Serge Hochar, who ran a winery outside Beirut called Chateau Musar, was alone in Lebanon. Years earlier he had sent his wife and three children to France so they wouldn't be killed in this war. He himself refused to leave the country. He walked to his window and looked out over the city, already shrouded in the rising dust and smoke of the attack.

The Syrians were aiming their massive ordnance at the Israeli-backed Christian militias in East Beirut. Ever since civil war had erupted in 1975, the city had been an orgy of hatred and shifting, incomprehensible alliances, Christians fighting the Palestinian Liberation Organization and Muslim militias, with the neighbouring countries of Israel and Syria constantly intervening and occasionally invading. Such a wreckage all these armies had made of the sparkling Mediterranean cosmos that had once been Beirut! They'd turned it into Beirut – international shorthand for the rubble of humanity that results when Jews and Muslims and Christians start taking sides and picking fights.

The Lebanese winemaker had lived and worked through this tragic mess since he was 35 years old. Now he was over 50. He was a man of medium stature, fit enough for his age,

---

Previous page: Charbel Abi-Ghanem, Serge's right-hand man who helped run operations at the Ghazir winery throughout the challenging war years.

though not physically imposing. His hair was grey. He dressed and groomed himself like any European businessman. Nothing remarkable there. His face, though – there was something exceptional about that. Serge Hochar was a sensualist, a taster, and he had the lush, exaggerated features to prove it: fuller lips than the average man, wider nostrils, more vivid eyes, and considerably larger ears. It was as if the man's every sensory attribute had been intentionally overdesigned by his maker.

Elizabeth Gilbert delivers her 2009 TED talk on how to unlock 'Your Elusive Creative Genius'.

When the shells fell around him that day, Serge Hochar knew the drill: go down to the bomb shelter in the basement with his neighbours and wait it out. Certainly, his top-floor, centrally located apartment was about the worst place he could be in Beirut. But this morning he lingered. Soon his neighbours came pounding on his apartment door – Serge! To

I'm beginning to see her point.

'You will journey to Lebanon, and I will enter you into new rooms inside your mind,' he is telling me, this 64-year old man who speaks with such high fervour, right here in the sterile lobby of the Marriott Marquis. 'You will come to Lebanon, and I will take you into the heart of the country, where you will see this culture where humanism started. I will journey you into this place that is 7,000 years old. I will teach you that intuition is more important than intelligence. I will take you deep into the world of taste, the world of the senses, the world where life is lived at its maximum, and you will ask yourself: Can this be possible? And you will say to me: Serge! No! This is not a possible world!'

So yes, of course I will go. Of course I will fly straight into the Middle East the very next week, even though I don't know a thing about wine. Of course I will be lured by Serge Hochar into wild, vital Lebanon, where a plastic surgery of gleaming new construction is already covering acres of war wreckage; where sexy girls in Gucci stroll the boardwalk beside modest girls in veils; where Roman ruins stand across the street from Dunkin' Donuts; where Palestinians are in the ghettos and ski lodges are in the mountains; where Syrian military checkpoints and fragrant wild oregano spring up side by side along the same highways. Of course I will travel halfway across the world to taste Chateau Musar and try to understand why Serge Hochar was willing to risk his life through 16 years of war solely for the pleasure and passion of making it.

'ARE YOU ENJOYING the weather?' he asks me. 'We have 300 days a year like this in Lebanon.' I am standing outside the Chateau Musar winery, in the mountains outside Beirut, breathing in air that smells of lemons and lavender. The sunlight is seductive. Over my left shoulder is a long view to the Mediterranean. It's lovely here. I wouldn't mind lingering, but Serge Hochar insists on hustling me down to his wine cellars – five floors below the earth, secure as a bunker, dark and silent – for the tasting that awaits us.

'You will taste your first Chateau Musar in the cellars of Chateau Musar,' says Serge (whose surname is pronounced hoe-shar). 'This is the proper way to do it. You will do many things in your life, but how often do you do something in the proper way?'

He has brought his oldest son, Gaston. Gaston is in his late 30s, quietly intelligent, modest in speech and dress, diligent with the winery's business, and a faithfully married family man. The difference between father and son is striking. There's Serge, with his wildly animated expressions, his high passions, and his French cologne. And Gaston, with his unassuming face, his neat American khakis, and his respectable reserve. Serge doesn't really get his son. Why so serious? 'He is his mother's child,' Serge says, trying to explain why this level -headed young man is the dead opposite of his sensual, reckless, Zorba-like father. A father who at this very moment is taking his first sip of wine and spitting it in a joyful arc over the rows of cob-webbed wine bottles sleeping silently in his cellar. 'I am baptizing them! The wine that is ageing in these bottles will taste my essence! These wines are my children! Now they will know that I am here!' Then, to me: 'And you will spit, too. Today you are not swallowing. Only tasting.'

He pours me some 1995 Chateau Musar. I make as if to taste, but he stops me.

Gaston and Serge: father and son with a passion for wine, part rational and sensible, part wild, uninhibited.

'No. First you smell. Then you will tell me what you smell.'

'Okay,' I say nervously. 'But remember, I've never studied wine before.'

'Good. Then nobody has ever taught you anything stupid and wrong. Now you smell.'

I shimmy the wine in its glass a bit. I close my eyes, bend my head, and breathe it in. And immediately it's not as if I'm smelling something; it's as if I'm seeing something – I'm seeing three distinct filaments of fragrance rising out of this glass at the same time. I can see them clearly in my mind, spiraling and braiding together like ghostly strands of scented DNA. It's beautiful, but I can't begin to think of how to talk about it. I don't want to lift my face from the glass.

'What do you smell?' Serge says.

'I smell blackberries. And winter vegetables, like turnips and beets. And rocks. And it's like they're all twisted together. Three-dimensionally.'

I glance up at him, checking in. He grins and nods, an encouraging sign, so I tilt my attention back down to the wine glass. Now I smell soil, big damp hunks of mineral-heavy, compost-rich soil, as if someone has just turned a clumpful of garden over with a pitchfork. It's gorgeous.

'This is amazing,' I say.

'Yes! Yes! This is what Lebanese wine is famous for – this aroma. There is nothing like this. It comes from the quality of the land, from what the French call *terroir*, the special richness of the earth. What is it about this land that gives such aroma? Is it chemical, biological, spiritual? There is something here, not like any other place. This is the biblical land of Canaan! This is the Garden of Eden! People say, "Serge, you are crazy – how can the whole history and archaeology of Lebanon be in your wine?" But do you see?'

I do see. And in trying to explain what I'm seeing, I start to veer off into the fanciful, unselfconscious, say-anything world where Serge lives all the time. It's easy to do this around him. He offers a kind of unconditional, universal permission to explore and express, which is why I now find myself saying: 'And there comes the smell of the blackberries again. And now there's a strong twist of something like gunpowder. Now the scent is getting darker and...horses! I smell big sweaty horses running through the woods! In the shade!'

Serge laughs and exclaims to his son: 'Do you see this, Gaston? This is what they mean when they say "Out of the mouth of the baby"!'

Gaston nods. Very polite. His mother's child.

But I am lost again in the floating, shifting aromas of this wine, a new sensation, a new evolution, every few moments. It is a long time before I can pull myself away from the glass to announce: 'Serge, this is the most exciting thing I've ever tasted.'

He gleefully corrects me: 'Aha! But you haven't even tasted it yet!'

FOR THE NEXT THREE DAYS, I put my senses entirely into Serge Hochar's hands. He oversees every morsel of food and drop of wine that goes into my mouth, all in the name of teaching me why the unique tastes of Lebanon have been worth defending through 16 years of war. I feel comfortable giving him this power over my palate, and he, too, is comfortable with this. In a fashion, this is what he's always doing as a winemaker: taking charge of other people's senses in this most intimate way. Drink me, his bottles demand on five continents. Feel this emotion I have felt, understand the history I have known...

In fact, Serge is accustomed to taking charge of other people in a good many realms. I can see immediately that he holds a rather godfather-like authority within his circle in Lebanon. 'I am not a businessman,' he says with a slightly dubious veneer of modesty. 'I am not a politician. I am a winemaker.' A winemaker who happens to preside over a small empire – the vineyard, the winery, the international distribution network for Chateau Musar, the luxury housing development and

country club he built in the mountains... Throughout the day, employees come to him for permission, neighbours seek him out clutching petitions for him to sign, and grandchildren are presented to him to be kissed and admired. He is met everywhere as a patriarch.

'Yes,' he admits: 'I speak to everyone as if I am the father, the authority, always. This I do on all levels. And I have no levels.'

Or almost none. In what is probably a very healthy turn of events, Serge's wife, Tania, looks to be the only person in Lebanon who remains undazzled by Serge Hochar. He married her when she was 20, a young girl of vitality and compassion. He, on the other hand, was a playboy – 'I was very much the freelance' – going out with four or five girls a night.

'For the first 10 years, she used to do everything I said, but only because she was still in awe of me. Then for the next 10 years, she fought with me all the time. 'Serge, I want a divorce! I want you to give me my freedom! You like the women too much!' But we do not divorce in my culture. So I give her the freedom she likes, whatever she likes, wherever she wants to go, however she wants to live. I know better than to ever try to tell her what to do anymore. Because what can a man do with his wife?'

Nothing like a long marriage to humble even a Zorba. Oh, yes, and also – Tania Hochar doesn't drink wine. She's allergic to it. 'Or so she says!' Serge exclaims. 'Sometimes I ask myself, Can this be possible? Can I truly be married to a woman who is allergic to wine? How could we be less compatible? We don't have any business to be married, but what can I do?' He roars out the biggest laugh I will ever hear out of him. 'This is what I call fate!'

Serge has decided that I should interview him at the Hochar family house, a massive place with an indoor swimming pool, an elevator, and terraces on several floors. It's far too big a house, Serge says, but it's what his wife wanted, and such things must be honoured. So we shall sit on one of the terraces, yes? And we shall drink a bottle of 1977 Chateau Musar while we talk, yes? We shall drink this wine slowly over several hours, so we can witness all its transformations, yes? This way he can study me, he says, even as I am supposed to be studying him.

'You have a natural nose,' he says. 'And when you drink wine I can see into your brain, so this is interesting to me. There is nothing as complex to the brain as smell and taste. When you look at something, you use only one part of your brain. When you touch something, you use only one part of your brain. But when you taste,

Seeking the higher ground – Gaston Hochar senior picked Mount Lebanon as a 'safer' place to make wine.

the entire brain is used because every memory of sight, touch, sound is needed to understand it. I am not a doctor, so I can only talk about the brain in an ordinary way, as a wild person. But I know about this, from my own experience. This is why, when I meet someone with an interesting brain like yours, I am excited. Because I would like to open up your brain.'

'Can you tell me about your family's history?' I ask, trying to shift attention from the rather graphic image of Mr Hochar opening up my brain. 'How did they get into wine?'

'But why do you want to know about such things?' Serge asks. He's already decanting the bottle of 1977.

'So I see into your brain.'

'Aha! So this is how it will go! *You* will ask the questions, and *I* will answer them!'

Well... yes.

'Okay,' he concedes. 'There are Hochars in Syria. There are Hochars in the northern regions of Tripoli. There are Greek Orthodox Hochars and Sunni Hochars. We are the Maronite Hochars. We are Catholics who have had many fightings with the Ottomans and the Shiites and the Palestinians throughout history. My family, it seems, has been in Lebanon for 800 years and in Beirut for 200 years. My father was raised in a family of bankers and traders, a very spoiled guy. He decided to go to France in 1928, and when he came back he started to be a winemaker. I was never able to understand the motivation of it.

'He planted some vineyards in the Beka'a Valley. But he put the winery here, just outside Beirut, on Mount Lebanon. He separated the vineyards and the winery because he knew that the Beka'a, which is on the Syrian border, was not a stable place. But he had to have a vineyard there because it is the best land, and there have been vineyards there for 2,000 years. But he wanted the winery to be safe here on Mount Lebanon, where the Maronites live. This is why you must drive more than an hour on the road to Damascus to get from the winery to the vineyard. This was always a dangerous drive for me during the war because whoever controls this road controls Lebanon; many armies have taken this road. During the war, I drove it many times with shells dropping all around me, and even once with the car right in front of me exploding and the driver killed.'

He sits for a moment without talking, solemn, one hand covering his face.

'I have tended to avoid talking about the war. It is abusing myself to go back to the subject again and again. I could tell you stories, but... you are asking me about my family. So. I will tell you that it was a scandal for my father to be a winemaker. This was something that farmers in Lebanon did, not the sons of bankers. But my father had this approach of class and noblesse about the wine. He was the first in Lebanon to put wine in bottles. Before, it was always in casks. He became the official supplier of wine to the French army when they were stationed in Lebanon, and this was a good business. He died when he was 62, very young, and I took over. This was easy for me, because he had given me responsibility to make all the wine already since I was only 18, because he could see that I had a gift, a philosophy of wine. I was always convinced to make a great Lebanese wine, with an aroma like the great wines of Lebanese history. And I was always a philosophic person. When I was only four years old, I was standing up and saying to my parents, 'We must only speak to each other with logic, for all of life is a question of logic.'

'What was your father like as a person?' I ask Serge.

'Well, he looked to details, and he was a person of nuance, but you are asking me questions about my father, and it has been over 30 years since he died!' Such a telling Serge Hochar moment: this man, a consummate purveyor of life, demurring on the subject of the dead. It's as if he can't even remember the dead, as if it is a physiological impossibility for anything associated with lifelessness to register in his brain.

'And between then and now we had a war, and it erased many memories,' he goes on. 'But why are you interested in this, anyway? This is something any winemaker could tell you – I am the son of this man, I was born in this year –

The Hochar family's original wine shop, opened by Gaston Senior in downtown Beirut in the 1930s.

but I am not interested in these things, and you should not be interested in these things. We should think only about the wine and how it makes your heart feel, why it takes you so close to God. So, here – now you may try the '77. It is ready for you to taste. And for today, you may swallow.'

THE '77 THAT WE ARE drinking was bottled two years after the civil war erupted. When the war began, Serge was only 35. He had taken control of Chateau Musar after his father's death and had his family and dozens of winery employees depending on him for security. He had no idea what to do. Revolution and war had upended the societies of Syria, Egypt, Iran and Iraq. Their elites had fled, never to return; many who'd stayed had been imprisoned or killed.

So Serge made a call on the oldest employee of the winery – a man in his 90s, a Mr Yousef, hired by his father decades earlier, respected by everyone in the community, and now sick on his deathbed. Serge said, 'Mr Yousef, please tell me, what do you think of this war? How long will it last?' The old man replied, 'Mr Serge, in Lebanon, when war starts, you can never guess how long it will last. But it usually goes on for more than 20 years.'

That evening, Serge called a meeting of the Chateau Musar employees. 'From now on, we are going to manage this company as if a war has started that will last 20 years,' he told them. 'We will freeze all the salaries, so nobody will get a raise. But nobody will get fired, either. We will find a way to reach foreign markets. Don't worry, we have enough stocks of wine to sell for 20 years. But we're going to continue to keep producing wine, too. Nobody will lose their job. We are staying, and we are not stopping.'

Shortly after the fighting began, Tania was in downtown Beirut when her car was stopped by a blockade of Muslim men with guns. They were going from car to car, pulling Christians out onto the street and executing them. A Muslim man leveled his machine gun at Serge's wife and said, 'Are you a Christian?' Tania said, 'Yes'. He looked at her face; she looked at his. And then for reasons nobody would ever know, the gunman said to his companions, 'No, not her,' and escorted Tania Hochar away from the massacre. He cleared a pathway on the sidewalk for her so she could drive past the roadblock, then told her exactly which route to take home to her family, in order to avoid more roadblocks. She had never seen him before, and she never saw him again.

'This', Serge offers once more, 'is what I call fate'.

Serge arranged to have his family escape from Lebanon. With the airports and roads under attack, the only way Tania and their three children could leave was in a small boat in the middle of the night, during heavy shelling of the port. They went first to Cyprus, then to Europe. Serge gave instructions to Tania that the children were to be educated to their utmost ability. Out of a population of three million, one million would flee the country during the war. More than 150,000 trapped behind in the chaos would be killed. Serge never considered leaving.

'I would never imagine to flee the country. I had to make the wine. If I wasn't – what is the word? – existential, I wouldn't have pursued it with such solemnness. Chateau Musar had to continue. It is an institution. I believe in

Gaston Hochar Senior (left) with one of his grandchildren and a visitor from France in the late 1960s.

institutions, like it is a religion to me. Making wine, making any art, is an institution, and if you don't keep it going you can't pass it along, and then it will die, and part of humanity will die. So how could I go to France, to California, and make wine? Now I shall tell you my flaws – ego and stubbornness and stupidity. How can I explain this? I am just an optimistic, idealistic, stubborn, stupid man. I felt that God wanted me to stay.'

As well it must have appeared, given the divine sphere of protection that seemed to surround the vineyard. In 16 years of war, not one Chateau Musar employee was killed, even as dozens of Serge's other friends and neighbours were lost in sniper attacks, car bombings, crossfire and bombardments. As for Serge, he continued to move through Lebanon as though bullet-proof. Driving alone one day to the Beirut airport to pick up a replacement part for his broken wine press, he found himself approaching a roadblock execution squad. Militiamen were dragging motorists from their cars and killing them. Not slowing down but not speeding up, Serge kept driving and continued on his errand. Apparently, he was just invisible that day.

The winery was never seriously damaged. The millions of bottles stored in its cellars slept through the years of conflict in silence. Serge even converted part of his wine cellar into a bomb shelter for refugees fleeing Beirut; it seemed the safest place in the country. And whenever the airports, roads or harbours opened, Serge shipped his precious cargo of Chateau Musar to a world wine audience that was increasingly turning its attention to this

Central Beirut 1976: by the late 1970s internal strife had torn this once beautiful and fashionable city apart.

vineyard that kept producing exquisite wines right from the centre of hell.

Production, of course, was always challenged. There was no Chateau Musar in 1976 because there was no harvest. There was total war – no fuel, no roads. The grapes died on the vine. In 1977, though, Serge again managed a harvest. He bottled a masterpiece blend of Cinsault, Carignan and Cabernet Sauvignon.

'I would work on the blend all day, then spend six, seven, eight hours a night on the terrace, looking up at the stars. Except, of course, when they were dropping bombs on me. This must affect the taste of the wine. Because wine is so complex, the winemaker has many options, and the option you choose in any moment is directly related to how you feel. If you are tired or stressed, or lonely, or under attack, this will undoubtedly show in the wine. There is no way to know how different this would taste if my family was not in exile or if there was no war. But at the end of the day, what only matters is this question: have you made a living thing out of this wine? Have you created a life force that is great and thriving? And can you accept the fate of how the wine turned out?'

Sitting on the terrace with Serge, looking out over glittering Beirut and the steadfast Mediterranean, we are working our way through a bottle of the very wine he produced in that fall of 1977. It is older and deeper than the '95 we tried yesterday. What the '95 did with aroma – shifting its shape and identity over time – this wine does with taste. It started off with a peppery, meaty flavour that reminded me of autumn stews, of pheasant and wild duck. An hour later, that savour was gone; the wine had become a spicy orchard. Then the smells deepened into something akin

Thick stone walls, plus a stubborn refusal to yield to danger all around, helped to keep Musar wine flowing.

to slow, low chords on an old piano. Now, at dusk, the wine has finally mellowed to the point where there is no more hint of fruit or food, only a deep, noble sadness. It tastes like a memory.

I report my taste experience to Serge, and he smiles with satisfaction.

'Yes. You are learning to recognize how the wine is changing over time. This is good.

'There is something happening in your brain right now, and I want to find the right word for it. You are being shunted, pushed from place to place within your memory. You do not even know all that is happening to you, and you will miss much of it, but the wine will forgive you for your mistakes. Wine will always give you more chances.

'If you are open to understanding change, this wine can teach you a lesson of tolerance. When you understand that all the flavours and smells and memories you have experienced over the last hours have come from the same wine, then you will learn not to condemn any wine until you have stayed with it through all its stages. When you understand that, then you can learn not to condemn any person until you have watched him through many stages, too. You even can learn to tolerate life through all its stages, never to make a complete judgement, always to know that there is more to be revealed. This is how taste can enhance your moral perceptions. I can see that you are ready to understand this. This makes me think you may be ready to understand Lebanon.'

HE FEEDS ME ALL OF Lebanon in three days. He brings me to the finest restaurants in downtown Beirut. In the small back room of a nondescript country grocery store near his vineyards, he presents to me 'the best sandwich in the Middle East,' and in another restaurant he orders me a dish that Lebanese shepherds have been living on for centuries: a

greasily delicious ground-lamb concoction served under a vividly yellow-yolked fried egg. He takes me to a humble eatery by the ocean, where we dine steadily for three hours. Serge hand-selects each course of my meal: buttery shrimp, eaten raw; tender young lobster; a tuna-like sashimi; and three kinds of fish that were alive only moments before, including one small trout-like specimen served to me fried and whole. Serge instructs me to eat the entirety of it, head, eyes, tail and fins. I lift the fish, but hesitate. 'Put it! Put it!' he commands, until I put it, as he exclaims: 'You are a beast! I mean this as a compliment.' Then he devours his own fish from tail to head and says: 'I am also a beast. I do not mind eating like the old man of the earth, like – what do you call it? – the humanoid, the caveman.'

In every restaurant, he is greeted with respect in English, French and Arabic, and with every meal there is Chateau Musar. Serge's wines are offered all over Lebanon as a kind of icon of national endurance. Their blended flavours perfectly reflect the blended culinary history of this country – exactly the kind of tastes, I suppose, that you would expect from a land that has been inhabited over the past 7,000 years by Phoenicians, Neo-Babylonians, Persians, Hebrews, Romans, Mamluks, Ottomans, Druze and the French. And tonight, our last together, Serge indulges me with an epic feast, celebrating over a groaning table the legacy left behind by this merging of world cultures. We eat tabbouleh, hummus, moussaka, halvah, baklava, mutton, olives, four kinds of bread, five kinds of cheese, stuffed grape leaves, herbed yogurt, fava beans, pomegran-

Lebanese vegetarian delicacies: perfectly in tune with today's healthy eating – and a glass of Chateau Musar.

Serge Hochar entertains: as in other areas of his life, dining with the master took on a larger-than-life quality.

ates, melons, cherries, dates, figs, apricots, grilled poultry, fish, lamb, fatted calf...

'Here, by a miracle of nature, you see that Lebanon still has its culture,' Serge announces. 'Even after 16 years of war and globalization of the world, even with the American fast food in Beirut, we are still eating this food that we have been eating since the time of the Bible. We are still a country of truth, something raw and pure. This is a miracle of cultural survival.'

Of course, Serge is a significant part of that miracle, having put himself at great personal risk to protect one of the pillar institutions of Lebanese culture. I don't know that this doesn't make him a hero of the world, too – even with all his 'ego and stubbornness and stupidity' – just for being a man who chose to protect something living during a time of wholesale death.

'Since I was young,' he says, 'I have always been confronted to try to understand that disturbing complexity called life. Now I believe that I have come to understand life through the wine. When I bring the yeast to the wine, when I see that union and its offspring, I can see that life – more than anything – is an eternal impulse to create. Life wants to live. Life wants to meet with life. Life wants to create more life. That is all there is. That is where the answers to all my questions are. That is where I can look for God.'

Serge Hochar long ago decided what he will participate in and what he will not participate in, and he stands hard by that. For instance, he often refrains from talking about the current American military presence in the Middle East. His brother and his son and his friends debate it constantly, but Serge tends to proffer that old metaphysical shrug of surrender when his opinion is sought. His attention instead remains on his personal fate. Which is, apparently, to continue making beautiful wine even as history boils up violently around him. Perhaps it is a gesture of defiance. Perhaps it is a gift to God. Or a gift from God. Or perhaps Serge is privately living out an old promise of peace from the ancient heart of the Middle East.

Look to the Old Testament book of Hosea. There, the prophet chastises his people for having fallen into iniquity and viciousness, for having stumbled off the path of righteousness. Amid the depravity and chaos, Hosea brings a simple offer of redemption from God: put your wickedness behind you, return to your faith, and you shall have new life.

'They that dwell under his shadow shall return,' goes the prophecy, that promise from long ago; 'they shall revive as the corn, and grow as the vine: the scent thereof shall be as the wine of Lebanon'.

First published as 'A Wine Worth Fighting For' by Elizabeth Gilbert in *GQ* magazine, September 2004.

Chateau Musar: these days, the Hochars' most famous vinous creation is very much part of the landscape.

WHEN HE PASSED AWAY, Elizabeth paid tribute to Serge with the following words:

'I want to try to honor Serge Hochar, the master behind Chateau Musar wines from Lebanon, which are celebrated worldwide for their extraordinary complexity. [The days I spent] with Serge were amongst the most epic of my life. He said to me: "When you encounter a bottle of wine, it is always a case of life meeting life. Especially if there is truth in the making of it. Watch how the wine changes, and watch how you change with it."

'He warned me against judging any wine too quickly. "We must let it grow," he said. "We must see what it becomes, because it is alive." Then he said something I've never forgotten. He said: "It is the same with people."

'He warned me never to judge people too quickly, because they are alive, and because life is always changing. People go through seasons in their lives, just like wines do. They shift, they grow, they move. Sometimes you have to wait it out and give a person time – to see who they become, and to see what fate does to them over the years. (Again, it is a case of "life meeting life", and anything can happen when life is in session.)

'I cannot tell you how many times I have thought of those words, especially when I am in conflict or crisis with someone, and I want to dismiss them or write them off. I think of Serge saying: "Wait. Don't judge them yet. It's too soon to tell. Life is happening. Let us wait and see what they become and what you become."

'He was a great, great man. Life meeting life, that's all there is.' 🏺

TAKING ON THE WORLD

How Chateau Musar crossed continents

S E R G E - I S M

I play poker only for pleasure, and I play games of chance to prove that
anything can happen at any time if only you trust, and if you allow yourself
to take risks and to sometimes be reckless. Do I play to make money? Why
would I do that? It is the same as with my winemaking:
I play to play, not to win money!'

SERGE HOCHAR

'LEBANON'S STORY IS FULL OF ups and downs, not like Switzerland's, which is sure and steady but very boring. We have had to take risks to survive,' explains Ronald Hochar.

The start of the Civil War in 1975 saw an almost complete evaporation of Chateau Musar's market: nobody was drinking wine – no one could afford to buy it. Soon enough, Serge and Ronald were forced to send their families to live out of the country, to Paris, in order for them to stay safe from the war and the shelling. They realized that the only way to sell their wines and to keep the family finances afloat was for these to leave the country too.

'Chateau Musar is not a wine you can send to a marketing man. It is something you have to deliver personally. So we decided to go to London, to tell our story, to show ourselves as the people behind the product,' remembers Ronald: 'This was a big risk. The easy option would have been to use an agency. We could have sold 1,000 bottles to Victoria Wine and this company would have distributed them for us, but we said no. How would anyone understand our wine this way? We decided that we would have more options if we stayed in control and told people about the wine ourselves.'

*Serge Hochar took Chateau Musar to London in the
1970s – merchants and restaurateurs were fascinated.*

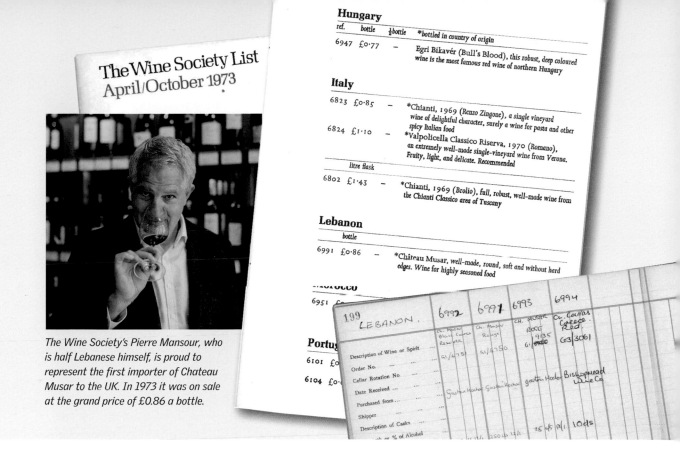

The Wine Society List April/October 1973

Hungary

ref.	bottle	½ bottle	*bottled in country of origin
6947	£0·77	–	Egri Bikavér (Bull's Blood), this robust, deep coloured wine is the most famous red wine of northern Hungary

Italy

6823	£0·85	–	*Chianti, 1969 (Renzo Zingone), a single vineyard wine of delightful character, surely a wine for pasta and other spicy Italian food
6824	£1·10	–	*Valpolicella Classico Riserva, 1970 (Romano), an extremely well-made single-vineyard wine from Verona. Fruity, light, and delicate. Recommended

litre flask

6802	£1·43	–	*Chianti, 1969 (Brolio), full, robust, well-made wine from the Chianti Classico area of Tuscany

Lebanon

bottle

6991	£0·86	–	*Château Musar, well-made, round, soft and without hard edges. Wine for highly seasoned food

The Wine Society's Pierre Mansour, who is half Lebanese himself, is proud to represent the first importer of Chateau Musar to the UK. In 1973 it was on sale at the grand price of £0.86 a bottle.

At first it was just Serge; he started working from his father-in-law's travel agency in Sloane Street. Slowly but surely he garnered interest from the local restaurant trade and more adventurous wine merchants. When war permitted, he'd go back home for the harvest, then return to London to keep spreading the word.

One of his first clients was the Wine Society in north London. Pierre Mansour, the Society's head of buying, proudly shows a ledger entry in a battered leather volume from April 1971, charting the purchase of cases from the 1967 vintage. 'Christopher Tatham was the buyer back then, when Serge and his father Gaston came to show their wines. He was dazzled: the wines were so unique, so different and unusual. They had that classic exotic twist that Musar still has, and they fitted with the Wine Society's objective so completely.' Pierre confirms that the Wine Society was the first to import Chateau Musar to the UK, and that it was the only company to do so directly. 'We only imported the wines directly ourselves on one occasion. Afterwards we went through Ronald when he was in the London office. The Lebanese know their systems: it's better to let them organize these things themselves,' says Pierre in a nod (not unfriendly, he is half Lebanese himself) to the specific organizational skills of his fellow countrymen.

The Wine Society has been a steady supplier of Musar in the UK since then. 'The wines have such amazing life, such vigour. We feel a great loyalty to them,' emphasizes Pierre.

But times weren't easy back in the 1970s. As Ronald says: 'We struggled to pay back our overheads at the start. We couldn't believe that we were shipping both wine and money in the same direction. So Serge said to me, "You go to London yourself, then you can see what is feasible".' Ronald could immediately see why Serge wanted to be in the centre of town – this was where the buzz was, where the buyers and restaurateurs (even in those Liebfraumilch-fueled days) were hungry for new and interesting wines. But the costs of living, rent, transport and bills were prohibitive, and the sales figures simply couldn't keep up.

In 1979, Serge was invited to take his wines to the Bristol World Wine Fair, where they were famously discovered by two prominent wine journalists, first Michael Broadbent, then Roger Voss. Broadbent, who headed the Wine Department at the prestigious auction house Christie's, called Chateau Musar 'the find of the Fair', writing a glowig report of it in the next edition of *Decanter* magazine. He said of the 1967 that it had 'a touch of claret, a bit of burgundy' and called it a great wine (a brave statement given the war-torn nature of the wine's origins). Broadbent invited Serge to the Christie's offices so that he could re-taste the

wines and be sure of their brilliance. Again, he was delighted, and widely said so.

Roger Voss was similarly struck by what he found. 'As I made my way around the fair at Bristol, I discovered a quiet booth tucked in a corner. A small, bright-eyed man stood there, but at the time nobody was paying him any attention. So I went up to him and we began talking. As we spoke, I realized that I was in conversation with a very serious winemaker.

'He had a quiet, unassuming way about him. Not showy or "New World" in style. He was just very passionate about what he did. He'd speak about the Beka'a Valley, the blend, and the war, then he'd step back and let me taste and make my own impressions. Two things stood out for me,' highlights Roger: 'Firstly, the shear quality of the fruit that had been used, and secondly, the way the wine had aged so superbly. Serge kept opening more and more vintages, telling me about them as he did so.'

Serge and Roger struck up a long friendship that lingered fondly through many long lunches, during which Roger learned about Serge's passion for all things French and his

---

The way we were: a snapshot of the 1979 Bristol Wine Fair (left) and Michael Broadbent officiating at Christie's.

Wine writer Roger Voss, who, along with Michael Broadbent, was an early evangelist for Musar in the UK.

offices were shared with the UK headquarters of Australia's Rosemount Estate. Ronald looks back with amusement at this time, as, at first, Lebanon outshone Australia: 'For a long time, we were by far the busier company. Lebanon was more popular than Australia. But then of course Chardonnay sales went mad and the Aussies went up, up, up!'

During the 1980s, Serge and Ronald worked hard at their UK sales and were richly rewarded by increasing customer loyalty. Bob Marshall helped as a sales consultant from 1979, and Jane Sowter joined Musar's own sales team in 1988; they both found they had a relatively easy job on their hands. 'It just happened to be a very well made wine, with great viticulture and fabulous winemaking. You could just taste that,' remembers Bob. 'When we first showed it at wine fairs, I think people felt sorry for me because I only had wine from Lebanon and some from Australia. But then they tasted it and raved, which didn't do sales any harm!'

Helped by reviews from Michael Broadbent in *Decanter*, Roger Voss in *Caterer and Hotel Keeper* and *The Wine Enthusiast*, plus the attentions of numerous Masters of Wine attending the wine fairs on the look out for new and intriguing wines, Musar began to catch attention. 'Consumer society was changing in the UK,' says Bob: 'As easy-drinking wines with English wine labels from the Southern Hemisphere began to flood into the supermarkets, increased disposable income and bankers' bonuses meant that customers really woke up to wine and could afford to start experimenting with what they drank. There was an enormous amount of interest. Then, of course, Chardonnay hit the big time and began to take over. But Musar was quirky and unusual – and it was there first.'

unique approach to blending Cabernet Sauvignon and Cinsault. 'He was a Francophile, of that there was no doubt. He wanted to use French grapes, but knew that it was the hot country varieties that succeeded really well in Lebanon. For him, the obvious choice was to blend Cabernet and Cinsault. In this he was way ahead of his time – he got to the Bordeaux-Rhône blend way before the Australians arrived with their "Cab-Shiraz".'

In 1982, Ronald made the decision to move Chateau Musar's office to a larger but more affordable building just outside London, in West Byfleet. The location was less prestigious but with its better access to the UK's major road and rail networks, the wines were suddenly within easy reach of Musar customers. Transport – and life in general – became far simpler. To optimize costs and space, the

The feedback was that Musar was complex and interesting, it appealed to the restaurant trade and private buyers alike, from the south right up to Edinburgh and Aberdeen. It could be Bordeaux-style or Rhône-like depending on the vintage. Bob recalls starting with the 1972 and 1973 vintages, then the 1977 ('which was fabulous') at only £6 a bottle. 'It was great to have an iconic wine – as that's what it was becoming – that didn't cost thousands. Musar was beautiful, and back then it was not expensive.'

The vintages of 1980 through to 1984 were years during which it was almost impossible to produce wine; 1985 was little better. Serge called 1982 'a pure wine of War', and the hardships he went through to keep his winery operating, his staff employed and his wine flowing, were near insurmountable. But he was determined not to fail. He now had a following, and refused to see customers go without his wine.

In February 1984, Serge received one of the greatest accolades of his life. *Decanter* magazine created a new award, 'Man of the Year', in his honour (awarded every year since, to luminaries of the wine world, male and female, for services to wine). As *Decanter* reported:

'In 1983, 3,000 people were killed in the Lebanon. The Chateau Musar vineyards are situated in the Beka'a Valley on the edge of the Chouf Mountains, scene of some of the worst fighting. Usually there is only a short journey from the vineyards to the chateau. In 1983 the Syrian lines ran through the estate so the grapes had to be taken on a dangerous 100-mile journey lasting five hours to reach the winery.

'So to introduce this new Award we unhesitatingly choose Serge Hochar as Decanter *magazine's Man of the year 1984 for his services to wine – and to the people of Lebanon. No one in the world of wine can have had such an appallingly difficult and*

The striking February 1984 Decanter *cover proclaiming Serge Hochar as the magazine's first 'Man of the Year'.*

dangerous job – and surmounted it. To produce wine at all in the circumstances would be remarkable, but to produce excellent wine is extraordinary.'

For Serge this meant international recognition. The impact of the *Decanter* award was tremendous. The magazine had a worldwide following and while Musar had been flagged up there by Michael Broadbent and his fellow writers previously, this was a personal recognition that would see him become a statesman alongside the most revered winemakers of all.

Serge was, of course, typically modest in his receipt. As Ronald remarks: 'Serge was not the type to show his ego, he preferred to have great comments about his wine. But the recognition he gained for Lebanon and the heroic nature of what he had done made him very happy. For him it was like a vinous Nobel Prize.

The years of stardom

Jane Sowter visited Lebanon soon after she began working at Musar's UK office in 1988 (Ronald employed her for her 'honest face'). 'Serge and Ronald wanted me to meet the team and visit the places they had told me about, to enhance my understanding of Lebanon. The Civil War had just ended. I was just a young thing from Norfolk, so this first visit was certainly an eye-opener, but it was also the start of a very long love affair with Musar.' Jane found the Hochars' commitment to their wine infectious. 'For Serge, it was running through his veins. After speaking to him, you'd leave the room wanting to collect every wine the Hochars had ever made. He knew that he had a jewel of a wine and wanted to share it.

'This was at about the same time as Tarek started at Musar and it was as if he and I were cousins. We learned the ropes together, with Serge teaching us. He would only allow us to do small jobs at first, just little by little. It was like a gradual coaxing, slow and steady. He needed to trust us before we were free to work with his precious product. He was just so in tune with his wines, he wanted to make sure we were too, so that we would convey the same message he did.

'It took 10 years for Serge to trust Tarek to make a wine on his own. It took about the same amount of time for him to give me my unofficial title: "My man in the UK". He was joking of course.

'After the war years in Lebanon, the Hochar family naturally wanted a secure and solid base from which to renew their markets. I was proud to be a part of building this platform with the development of Chateau Musar UK,' says Jane. She and her team gradually expanded sales beyond their favourite niche wine outlets – and reliable sales into Ireland via Ronald's contacts at Grants of Ireland (now Findlaters) – and developed a larger following in high-end stores such as Harrods, Selfridges and Waitrose, as well as other independent

retailers. Serge relished the fact that the UK was a very open market; being a 'non wine producer' he found that customers there were keen to embrace 'unusual' Lebanon. It was the same with Holland. He and Jane developed a lasting relationship with importers Cornelius Dolle and Lex Jongsma, who really took Musar to heart. This was a country with a vibrant fine-dining scene and it needed the wines to match.

The 1990s saw expansion across the UK, further into Europe and then into Scandinavia, an important Musar market today.

'In Sweden, Norway and even Finland, they loved the taste profile of Musar,' explains Jane. 'Its rich, opulent style seemed to complement

A single-vineyard triumph: 'Hochar Père & Fils', aka HPF, or as Serge eventually conceded 'Happy, Positive, Funny'.

their cuisine, climate and lifestyle. We found that there'd be dining clubs and collectors who looked forward to every new vintage. And there was always a running joke that if an older wine had sold out, we'd always find it in a collector's cellar in Stockholm!' Jane goes on to describe a mini-revolution when Musar's Norwegian retailer decided to delist the wine: 'People wrote and wrote again to complain, and eventually they had to reinstate it. The pressure of people-power was so great!'

The expansion program also saw a new label developed, 'Hochar Père & Fils', in 1989; a brand that Chris Murphy of Marks & Spencer spotted and decided to make the cornerstone of his new 'Seven Winemakers' range. (It was

Watching over them: the statue at the winery's shrine to Saint Rita, the patroness of impossible causes.

released alongside prestigious wines from Antinori in Italy, Mondavi in California and Len Evans of Australia.) 'This was a new concept range' explains Jane: 'It was meant to last for about nine months but it was very identifiable and really, really successful. Once M&S had come to the end of their run with it, we had such a huge demand that we just continued making it. Now it is a classical and much loved wine, the only single vineyard wine in our portfolio, released three to four years after the harvest with a few months of oak ageing, we consider it a younger "Chateau".

'It exasperated Serge that he had created this new expressive, elegant wine and customers called it "HPF",' adds Jane: 'Why simplify it?' Marc laughs at this, adding: 'But in the end he gave in and added his own interpretation to anyone who'd listen: "Happy, Positive, Funny". He wouldn't take himself too seriously.'

Through its expanding export markets and increasingly loyal customers at home, Chateau Musar UK was fast becoming the largest outlet for the Hochars' wines, and the driving force behind global sales. (Today, Chateau Musar UK, with its office of seven staff, distributes close to 70% of all Musar wines.) 'Serge loved seeing these worlds colliding over his wine,' remembers Jane: 'And when we started selling in Canada, he was proudest of all. It was tough to get through all its liquor board regulations and lab-testing at first but we met an importer called Jack Segal in Alberta who helped us. He and Serge became fast friends and used to buy each other expensive cigars, which they'd smoke together and put the world to rights.'

The 1990s' age of experimentation saw a booming interest in wine triggered by new ventures *Wine* magazine and the International Wine Challenge which had huge success in

Thoughts on Musar and Serge

by Sarah Kemp, publishing director of *Decanter* magazine 1996–2017

I FIRST CAME UPON Chateau Musar in a small corner shop owned by a Lebanese gentleman in Paddington in the late 1970s. No fine wine shop, just a local corner store; I bought a bottle, it was my first fine wine, and I can still remember today its exotic and mesmerizing taste. From corner shop to world stage, it's hard to imagine that 40 years later Musar would be so feted that it would command its own masterclasses in London, New York and Shanghai, sitting happily on the programme next to Bordeaux's most illustrious estates. But maybe not so hard, because even then it was evident that Musar was different. It was also run by a very different producer.

I knew Serge for over 30 years and watched with admiration as he introduced his wine to increasingly fascinated wine lovers. Philosopher, humanist, naturalist, he was the rare beast who saw beyond fruit flavours and points, his belief was that authenticity was the greatest beauty of all. For some, the lack of homogenization was a challenge, especially during the 1980s when the style for squeaky-clean extracted wines was at its height, but Serge knew, and proudly stood by his wines, knowing that they expressed their ancient *terroir* with a consistent voice.

Serge, philosopher, part Gandalf, part Ariel, proudly told me that the first evidence of trading of wine by the Phoenicians was in Lebanon's Beirut museum – this link to the past is very much part of Musar's DNA. The wines have a spiritual as well as an intellectual dimension, and like a fascinating discussion they continue to occupy one's thoughts long after the last sip.

Serge was particularly proud of his white wines. On the 30th anniversary of the *Decanter* Man or Woman of the Year Award being founded (he was the first person to receive the award, the story even made the BBC news) Serge generously offered to throw a dinner for his fellow recipients. Amongst those present were Angelo Gaja, May-Eliane de Lencquesaing, Jean-Michel Cazes, Paul Symington, Ernst Loosen, Piero Antinori, Hugh Johnson and Michael Broadbent. Great wines were served, 2000 Château Lynch Bages, 1992 Ridge Monte Bello as well as 1995 and 1972 Chateau Musar red. The last wine served before the port was the 1969 Chateau Musar white. It was the talk of the night, with its cashmere texture, sensory complexity and length, which seemed never to end, as if we were hearing beautiful voices holding their notes until we could hear no more. Serge knew by putting it at the end of the flight of wines over dinner it would be compared to the great reds. I watched Serge's face: a quiet smile; he knew without doubt the point had been eloquently made.

making wine more accessible. Musar found its place here, too, and continued to thrive. Richard Hunt joined the UK office in 1997 and recalls new levels of excitement as collectors began to anticipate the next Musar release: 'Each year was different, and customers really wanted to see and taste what Serge would create next.'

After a steady run through the 1990s with its two lead brands, a younger-drinking Musar label 'Jeune' was created in 2007: an unoaked blend of Cinsault, Syrah and Cabernet from younger Beka'a Valley vines meant for drinking while fresh, young and fruity. Enormously popular, this then led to the development of another new style, 'Koraï', pitched between the Hochar Père & Fils and Jeune levels, and aimed at the younger audience in Lebanon who have a real interest in developing wine knowledge.

Tania and Marc

There's little doubt that, with the arrival of 'Jeune' and worldwide acclamation for the wines and the family, Musar's journey to stardom was complete, but its upward trajectory wouldn't have continued in the way it has without the presence of two more characters in its story, Tania and Marc.

Tania Hochar, Serge's wife, never did drink wine. It does not agree with her. But she has been the reassuring presence that kept the family together during the tentative first steps of winemaking, the marketing forays into Europe and the UK, the brutal upheavals of the civil war, the loss of her husband and advent of the new generation, not to mention the politically turbulent times that have peppered the country's history since 1990.

Gaston and Marc admire the masonry at 'Chateau Mzar' as the old convent is prepared for a convivial future.

Tania is a cook and food lover; she matches Serge's ingenuity with the meals she creates, but doesn't have his tempestuous nature. She instinctively knows the Lebanese dishes that will show Musar wines at their best – although her glass will simply contain water. It was Tania who kept Gaston, Marc and their sister Karine safe in Paris while Serge was dodging the tanks to make his wine. And it was Tania, in the 34-day Hezbollah war of 2006, who sewed the Musar crest onto bed-sheets, so that they could be spread over the trucks to protect them from the fighting as they journeyed to and from the Beka'a with the harvest. Tania is the level-headed one who dispenses calm, where all about her are at battle and quest.

And it is Marc Hochar who has injected so much passion and energy into the company since he returned to the family fold in 2010; who has kept up the propulsion begun by Serge in expanding Musar's sales, and who has ensured that the Hochar legacy not only continues but profits, too, in order to invest in the winery and expand its reach.

At the start of his career in finance, Marc made several trips to Asia that awoke a passion for the region. 'For me it has similarities to the Middle East. I love its food, the difference in its cultures, the hunger to work and eagerness to discover new things. In 2010, I felt Asia was on the verge of an explosion of interest for absolutely everything – including wine – and that we needed to be there to be part of a new adventure. It could only help us grow. I also realized that there were many countries without importers – Taiwan, Japan, Vietnam, Singapore and more – and that we hadn't visited our existing importers to explain our wines and our philosophy, how to serve and taste them. Our customers were working blind and

Broadbent père et fils. *Bartholomew was instrumental in introducing his father's 'discovery' to the USA.*

just with their palates and intellects but with their souls, pushing everyone to slow down and to dig deeper, to taste not merely the product of vines but the product of civilization. '[He was] a winemaker who made you think differently not just about his wine, but about the world,' says Paul Grieco, the owner of New York City's Terroir restaurant. 'You couldn't attend a tasting with Serge, walk away, and not feel differently. It was impossible.'

I MET SERGE Hochar in 2001, almost immediately after moving from Mississippi to New York to work for Broadbent Selections, which had just started importing Chateau Musar the previous year. The wines had been available in the United States since 1973, when a Lebanese-American businessman named Richard Traboulsi began importing Musar. In 1981, shortly after the wine critic Terry Robards, in *The New York Times*, called the 1970 Chateau Musar Rouge 'astonishing' (in a blind tasting, he'd guessed it to be a 1970 or 1974 Cabernet

from Napa), the Hochars – Serge and his brother Ronald, who handled the business side – moved to Négociants USA, and then later, briefly, to Remy Martin Premier Wines. The move to Broadbent Selections made a kind of historical sense. Chateau Musar's entry into the international spotlight, after all, had come at the hands of Michael Broadbent, who'd proclaimed Chateau Musar the 'discovery' of the 1979 Bristol World Wine Fair. Serge saw Broadbent Selections, founded in 1996 by Michael's son Bartholomew, as his conduit for re-introducing his wines to the US market, for garnering the sort of rapt reverence that Michael Broadbent had spurred in the United Kingdom. I was fortunate to land at the heart of his plans, and to have Serge Hochar, more than anyone else, guide me into the wine world.

Chateau Musar's red wines, at the time, were what sommeliers and consumers knew best from the estate. The whites had a mixed following, but the quantities available in the States were very small. A few sommeliers – Paul Grieco and Il Buco's Roberto Paris among them – were early devotees of Serge and his wines. Serge and I decided, after lots of slow planning, to put together a vertical tasting for the New York City trade. The idea was to showcase older vintages of Musar with the hope that the industry would pay attention in a different way, the way Grieco and others were.

Serge had poured older wines before, in certain settings, but never quite like this. We served six reds and six whites, one vintage each from the 1950s, 1960s, 1970s and 1980s along with two vintages from the 1990s. Serge insisted that we serve the red wines before the

white wines, the youngest wines before the oldest wines, and that we double-decant every bottle one hour in advance, and that the attendees commit to a two-hour seminar.

Our expectations weren't lofty; we hoped to educate, enlighten and just maybe to inspire a few sommeliers. The response, however, was electric. At the end of the tasting, sommelier Robert Bohr was so moved by the 1954 Chateau Musar white that he purchased what few bottles we had to pour by the glass at the restaurant Cru. That first tasting seeded a small but significant cult following, especially for Serge's white wines – which he cheekily called his 'first red wines' and which have inspired competing waves of love and loathing across the country.

'Musar is probably one of the most important wines of our post-modern wine industry,' says Pascaline Lepeltier MS, managing partner at New York City's Racine restaurant. 'What Serge Hochar accomplished with his work and his passion had a tremendous impact on a lot of us sommeliers, especially in this crucial time between 1980 and 2010. We'd become convinced that great wines could only come

FROM BEKA'A TO BEIJING

by Edward Ragg MW and Fongyee Walker MW

OUR FIRST EXPOSURE to Lebanese wine was, as for many wine lovers, with Chateau Musar. Unsurprisingly, considering Musar's initial growth outside Lebanon, this occurred in the UK. Ever since Michael Broadbent had 'discovered' it at the 1979 Bristol Wine Fair, the winery's international star had deservedly risen.

Our first experiences of Musar took place at fabled Oxford Lebanese restaurant Al Shami. Dining here gave the perfect opportunity to experience the evolution of red Musar as older vintages of the Chateau blend and the more approachable 'Hochar Père & Fils' label were readily available at (given the quality) especially favourable prices. And what better way to enjoy Musar than with Lebanese fare?

In a more formal setting, at Cambridge Wine Merchants, where Brett Turner invited Ralph Hochar to present his family's wines,

No longer obsessed with super-premium Bordeaux, the Shanghai restaurant scene now casts its wine net widely.

Fongyee's conversion to Musar white was almost instantaneous, whilst Edward took the chance to seek out older vintages of Musar red wherever he could.

Looking back at Edward's 'wine diary' from 2005 we see that Ralph chose to show the reds before the whites, and from younger to older: both wise moves with the particular Musar vintages on offer. Of the reds, the 1998, 1995, 1989, 1979 and 1972 were present. Of these, the stand-out vintages on the night were 1995 and 1979. The rich, layered and aldehydic whites of the 1998, 1994, 1989 and 1969 vintages followed. Of these, the 1989 and (especially) the 1969 were fabulous.

The whites – which in retrospect, may have anticipated the 'natural wine' movement – found a passionate fan in Fongyee with their textured complexity and capacity to age. In fact, that tasting demonstrated not only the overall ageability of Musar, notwithstanding vintage variation, but also Musar's idiosyncratic approach to winemaking in a world that was increasingly shaped by the overall improvements of wine technology of the 1990s and thereafter. If you wanted zip, bright fruit, micro-oxed tannins – indeed, plenty of polish all round – Musar was anything but...

Arguably, white is still Musar's more controversial colour. However, given the penchant some adventurous fine wine lovers have recently shown for traditional white Rioja – most obviously Lopez de Heredia's wines – it is perhaps more understandable that Musar's white should have found a new following in some parts of the world, including China.

Partners in wine: Fongyee Walker MW and Edward Ragg MW set up Dragon Phoenix Wine Consulting in Beijing in 2007 in anticipation of the surge in popularity of fine wines in China over the last decade. Both of them are long-time fans of Chateau Musar.

WHEN WE SUBSEQUENTLY moved to Beijing in early 2007 to establish the wine education and consulting business that became Dragon Phoenix Wine Consulting, we could have been forgiven for feeling nostalgic about the UK wine scene we'd left behind, especially given the ease with which we could find distinctive wines like those of Musar. Not only was the 48% tax on all imported wine a sobering consideration, the margins levied by Chinese importers, distributors and retailers were jaw-dropping, propping up some hefty salaries in the larger import companies that controlled the market. Added to which, there was a lack of diversity beyond the typical selections from France and Australia.

The mainland Chinese market in which Musar would eventually carve a niche was distinctly polarized back then. The super-premium wine of choice was Lafite – 2008 saw the heyday of the 'La Fei' craze – followed by similarly prestigious *grands crus classés* and a cluster of other iconic wines from around the world (Penfolds Bin 707 and high-end Napa, for example) at one end of the scale. At the other, was a veritable sea of bulk wine of markedly different quality levels. Whilst there were some mid-market wines, these represented a small part of most importers' fledging lists. One of us even remarked morosely (as it turned out, incorrectly): 'We'll never drink decent wine again!' Which serves to illustrate how undeveloped the mainland Chinese wine market was at the time of the Beijing Olympics. Not until after the Olympics, in fact, did China begin to make a larger impression on the international wine industry.

Musar made its first moves into Asia back in 1999, working with FICO in Hong Kong. The former British territory no doubt still held some ex-pat fine wine lovers with a soft spot for the Lebanese icon. Certainly, Musar's success in the UK gave it a sure footing for entering Hong Kong at that time. Its recognition was also supported by itinerant sommeliers from Europe who were a point of connection with the nascent home-grown wine service soon to show its colours. When Hong Kong unexpectedly removed tax on imported wine in 2008 – in the hope of making the territory an 'Asian wine hub' of sorts – Musar increased the availability of its vintages in the region. This

was a popular move. To this day, Marc Hochar reports that Hong Kong is still Chateau Musar's most important Asian market with the widest range of its wines available.

But what of mainland China? Marc had joined the family business specifically tasked with developing the winery's international and potential Asian markets, and he began working with importer Summergate in 2010. Summergate was notable for its risk-taking: early on, it built up its domaine burgundy list at a time when China's thirst for burgundy had yet to develop; it supported brands from countries with little recognition in China, and it led the way for New Zealand with Villa Maria – for many Chinese consumers their first taste of Kiwi wine. Summergate's strengths in supplying the developing Chinese on-trade gave Musar a chance to be listed in the right places, at least where professional wine lists were operating. Largely, this was in the first-tier cities of Beijing and Shanghai and, to a lesser extent, in the southern first-tiers of Guangzhou and Shenzhen.

With his father, Marc had routinely visited mainland China on 10-day trips, convening tastings for trade and private consumers, initially in Shanghai and Beijing, but increasingly further afield. Having gained some recognition in these major internal markets with Summergate, it was perhaps not surprising that Musar then chose to change importer in order to reach adventurous wine lovers in smaller cities. Saturation of the first-tier city markets, especially in the on-trade, was an unsurprising phenomenon that led many fine wine producers to do the same, from as early as 2012. By 2017,

Tang Zhu of Dragon Phoenix Wine Consulting hosts a fine wine masterclass at Prowine Shanghai for new-generation Chinese gastronomes.

Musar had partnered with east coast-based Healthy Fine Wine with a view to widening distribution through Musar-savvy wholesalers. This gave Musar brand recognition as the Lebanese fine wine of mainland China – albeit in a specific niche – and perhaps more importantly, as an unusual fine wine in its own right.

In our experience, it is the more adventurous Chinese wine lover who gravitates towards Chateau Musar. With classrooms in Beijing, Shanghai, Chengdu, Chongqing, Shenyang, Dalian and Qingdao, Dragon Phoenix has a network of students across the country that has given us first-hand experience of how Chinese wine loving palates have developed over the last 12 years or so. In Shanghai, Fongyee taught a specialist class on Musar which sold out in a matter of minutes and ended in a classroom of eager students tasting and discussing the wines back to the 1997 vintage. This classroom session, was, of course, followed up by a fine Shanghainese dinner with multiple bottles – the

Musar red, seen here maturing in French Nevers oak barrels, began making an impression in Asia in 1999.

savoury notes of aged Musar singing with the intensity of various braised dishes.

The food and wine culture in China is particularly special in Chengdu. Here, Sichuan cuisine is prized alongside a more laid back, leisure-orientated, tea-drinking lifestyle. In this particular market, adventurous wine lovers are increasingly common. Marc shared with us his surprise at how well Chateau Musar's 2004 white accompanied Chengdu hot pot. The savoury, nutty, aldehydic and saline character of this wine undoubtedly pairs well with chilli spice and the numbing character of Chengdu cuisine; whereas tannin in Musar red, even in an aged state, is more emphasized by chilli heat.

In practice, though, how well such a wine 'works' all depends on what the Chinese call 个人的口味 'ge ren de kou wei': your individual palate sensitivity and preference. Among our Chengdu students, some will inevitably prefer Musar red because it emphasizes 'heat' more, or perhaps prefer white Musar because it moderates it, depending on what they are eating. More than 'heat', the savoury flavours of Chengdu cuisine also lend themselves well to aged white Musar's nutty, honeyed notes.

China does not have a culture of matching specific alcoholic drinks, let alone wine, with food, but this lack of tradition clears the ground for considerable experimentation among Chinese wine lovers, all of whom love to eat!

One of the key factors supporting China's increased interest in 'eclectic wines' like Chateau Musar has been the influence of Chinese social media, first with Weibo and, more significantly, with WeChat. With the rise in wine education and a new generation of ardent wine lovers searching out new wines, WeChat, as the principal mode of digital communication, has fostered virtual groups for wine lovers and their friends, sharing tasting

Visitors taste wine at the 90th China Food & Drinks Fair in Chengdu, capital of China's Sichuan Province.

experiences, photos, videos and personal recommendations. Nor is this interaction only virtual. Those who record enjoying Musar on WeChat will almost certainly be doing so with wine-loving friends in Chinese restaurants. Communal eating and enjoyment of food culture are of such central importance to Chinese identity that any wine finding its way meaningfully to the Chinese dinner table is likely to get noticed. An introduction from a friend – whether it is to a new wine or to further business – has a unique value in Chinese thinking and everyday life.

We close with a further personal reflection. Just as Musar was getting off the ground in mainland China, Edward wrote a poem, entitled 'Chateau Musar', which appeared in his first book of poetry, *A Force That Takes*. The poem was prompted by an interview Serge Hochar gave to *Decanter* magazine. When asked about the difficulties of producing wine in war-torn Lebanon, Serge replied: 'We are inoculated to resist'. This nuanced statement combined the language of winemaking with the notion of political resistance; it implied the kind of resistance to conflict demonstrated by the Hochar family's determination in continuing to produce wine whatever the circumstances. (The poem is printed on page 202.)

Survival, consciously or otherwise, may be the basic aim of any life, but Chateau Musar, through the diligence and determination of the Hochar family, has truly delivered on Serge's original promise, after taking over winemaking from his father Gaston in 1959, of making Musar 'known worldwide'. From Beka'a to Beijing, through the virtual channels of Chinese social media and at the Chinese dinner tables of this country's advanced wine lovers, Musar's star has risen yet higher still. ✤

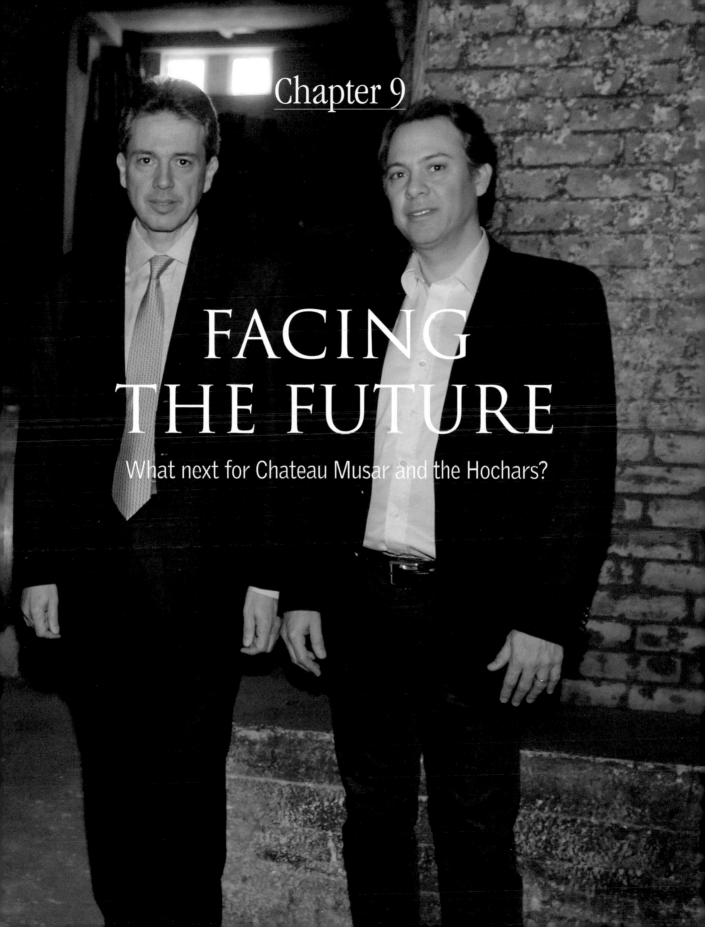

Chapter 9

FACING THE FUTURE

What next for Chateau Musar and the Hochars?

Chapter 9

T R U I S M

'The Hochar wines are our wines, and Serge's legacy is our pleasure. The family treasures, unearthed from the gargantuan Musar cellars, are ours to discover and enjoy. Each bottle is both the same and different, just like all of us.'

KEVIN GOULD

FOR THE HOCHAR family, the clock does not tick inexorably forwards. No matter how international, how multi-culturally aware they are, the family is before everything else, Lebanese. To be Lebanese, whether domiciled or diasporic, is to be acutely aware of the measure and cadences of time. When your culture spans millennia, when your myths and generations stretch back to the dawns of writing, of commerce, of the wine trade itself, your place and legacy in time becomes more apparent than the news you make today. So as much as Serge's clock was always ticking and never still, it also had a second hand that counted backwards to the beginning of the story that started with the Canaanites and the Phoenicians, a story that was ancient by the time of the touristic Greeks.

When a Hochar talks about length, he or she is not just describing the time that a wine dwells on the palate, or the endurance value that a certain bottle or vintage exhibits in continuing to evolve for decades after its maker has quit the scene. When a Hochar talks about length, there is an important back story of recorded time, and the family's inevitably modest place in it. In this case: they ask, are we worthy to be counted as successors to the first ever winemakers and traders? Do our wines honour their birthplace and its culture even as they honour the person who pours them?

In a way, this constant backwards glance – combined with a Lebanese confidence that comes from deep in their DNA – has kept the Hochar family immune to the diseases of fashionability. Not for them the St Vitus' dance of following the markets; with every rash of competitors the family doesn't rush to scratch the itch of change but reaches instead for the emollient balm of time. Fashions come and go, but style is always in fashion.

For this family and its wines, their ability to reach out beyond the horizons past and present has made them peerless. And by refusing to see everything that they do through the prism of 'now', the Hochars ensure that their reputation and legacy is also timeless.

For example, the great fashion today is for 'natural wines', made with as little interference from man as possible. The Hochar family has never taken the opportunity to re-label its wines as 'natural', even though it is amongst the first in the modern era to be entitled to do

so. It does not whip up a frenzy of Instagram followers to whisper that it is the grand-daddy of the 'natural' movement (even though it is), but it continues to make wines the way that it has always done: the way that suits the grapes, and the conditions, their culture, their taste and their heritage the most exactly.

And nor are Hochar wines historical set pieces that grope along following the late master's vision. Although Serge has gone, Ronald, Gaston, Marc, Ralph and Tarek, and the Hochar teams in Lebanon, Europe, the Americas and Asia, are still energized by the inextinguishable enthusiasm that he breathed into them. Their belief in the unique character of each sip of their wine, and its power to transport through the human history of the vivacious, unsettled Middle East to our own culture and our own time is as vital to them as it was to Serge. The belief is that every glass from every bottle is there to be tasted and savoured by everyone.

'The family doesn't rush to scratch the itch of change', and as its clock ticks forwards (and backwards), the Hochars still embrace the quirks of their original winery.

Each bottle is both the same and different, just like all of us. In every sip are both memories of the past, and memories made for the future...

Ronald's way for the future

Serge readily admitted that he was a gambler, he liked to take a chance on his wines. He would give them every opportunity to express themselves without his intervention. But his approach was also highly academic: he stood fast by his winemaking beliefs and theories no matter what the practical cost. This was where his brother came in. As Musar grew into a successful business, Ronald was more than just a steadying hand at the financial helm. He offered the counterbalance to Serge's enthusiasms, and found the middle way between risk and idealism. Serge would say: 'We must sell in Europe!' Ronald: 'Let's start with France.' Serge wanted to experiment with more new oak, and Ronald would counter with 'OK, let's go up by 10%'; Serge: 'But you must let me try!'

'We had the same philosophy, but a different approach to the real world,' says Ronald: 'It was not always easy, and we had to make concessions to each other, but we were two brothers with the same goal, and the times we lived through forced us to be strong together.'

For Ronald, his role today is to transfer the flame from one generation to the next. 'Gaston

Ronald stands by the well where he and Serge, as boys, washed wine bottles one cold winter: his greatest wish is for Gaston and Marc to have the same fraternal bond.

and Marc: I wonder if their characters are the same as mine and Serge's? Will they take the kind of risks we did? Would they react the same way to what we experienced? I don't know. But what I do know is that they have the same family spirit. They have tolerance and they have open-mindedness. That is what is so important. My job is to keep that in place – for them, and for the next generation.'

Gaston: new developments

For Gaston it is essential to maintain the strong grape identity of Musar's wines. This is something he forged alongside Serge, and that he is determined to build on. 'For Chateau Musar we have Carignan as the differentiating grape in the blend; for Hochar Père & Fils it is Grenache; for 'Jeune' it's Syrah, and for the rosé, Mourvèdre.' These feature grapes aren't bottled individually (which is perhaps a point of contention with Tarek) but act as a flavour guide and a major point of recognition.

Gaston sees this clarity as a vital tool for customers as they get to understand Musar wines and are guided by the distinctive characters these grapes gain in Lebanon. The vineyards tended by Musar are – and always have been – organic, which ensures the' flavours of their grapes are accentuated and vibrant. 'This is the Musar way. Our vineyards are certified organic, but this is not something we use as a selling point,' says Gaston with calm precision: 'It is more important for us that people respect our wines for the authenticity they have, and the way they represent the best of our *terroir*.'

Chateau Musar does not have a grand programme of expansion in place. The team is

Gaston's vision for the future sees continued recognition for Musar's white wine – as his father would have wished.

carefully trying to increase its vineyard area, but it is proving hard to do so. 'When we can buy a small plot or a new vineyard in the Beka'a where we understand the *terroir*, we do so. But this is very challenging as in Lebanon being discreet is not an option. Everybody knows everyone else's business. Once people realize that we are the ones buying the land, prices goes right up. A lot of subtle negotiation is needed!'

Tarek found a new plot of land with great potential in a new area for Musar near Byblos on the sea-facing side of Mount Lebanon. Ronald and Gaston agreed to develop it to test a new *terroir* and expand the regions of Musar's vineyards. 'We have found a plot that we are gradually clearing to make it ready. It is limestone based, with white and yellow soils. One day we think it will be great for the white grapes, but it is difficult to plant as it is very rocky.' The rocks, as with everything else at Musar, are recycled. 'The small ones are broken up as we plough, and the bigger ones are piled

For Gaston, Chateau Musar will always be about organic winemaking, no matter what the challenges climate change may bring.

up around the sides of the vineyards as a wall. This saves money on construction, and by using what we already have, we make it sustainable – the boulders are already part of the scenery.'

'We don't like things to be artificial,' he continues: 'Our fertilizers are not artificial either; we use the seeds and grape skins left over after the wine has been fermented and pressed so that we give back to the vineyards and fields what we have taken out.'

Adapting to new climatic conditions is a worry for the future. 'There is no doubt that they are changing,' says Gaston: 'September is our harvest time in a normal year, but lately we have started bringing in our Chardonnay and Viognier in early August. And usually we have two or three storms of rain by November, but this year [in 2019] we have had nothing. Until now, we haven't struggled for water. At the moment we are OK. But the big question is until when? We may have a lot of rain all at once, but this is not what anyone wants as it causes erosion and flooding.'

Gaston works closely with Tarek in the vineyards. Over the years, Tarek has proven himself to be a fine farmer and a brilliant viticulturist. He made his way by being a diligent, intelligent listener – first listening to Serge and getting to know his way with wine, taking in every implication and ambiguity, then to the land and to the vines themselves, recognizing them as fellow communicating creatures. Serge learned to trust Tarek, calling him 'the grape guy', giving him a free hand to plant and cultivate the grapes that would best honour the Hochar name.

Tarek is keen to do more work with Musar's individual grape varieties, bottling them young, separately and without oak, to show their unique Lebanese expressions. Gaston is char-

Tarek seeks a way forward for single-varietal Musar wines that show Lebanon's unique grape expression.

mine its identity. 'Our blend is done in a very specific way, after ageing in casks. It is a very individual process and can't be copied. We don't need to change it, only adapt to new vintages and new weather conditions.'

If he shows just a little of his father's stubbornness here, it is not without good reason. 'Other wineries have just started selling Obaideh and Merwah as mono-varietal wines, so we don't need to. There are two Obaidehs and two or three Merwahs. This is a really good thing as it creates awareness of the grapes. Then people will try our white to see what the two can really achieve when they are together!'

Only proving Merwah's heritage would persuade Gaston differently. The Hochars would like to ascertain if Merwah is the ancestor grape of Sémillon (famously the mainstay of Bordeaux' great Sauternes wines and one of Australia's finest heritage varieties), but they aren't 100% sure. 'Carrying out the proper DNA trials is very expensive and the State has no funds to support this kind of research.' Gaston believes the familiarity of the name Sémillon would help sales, but alternatively, with the growing global interest in indigenous grapes, having Merwah remain uniquely Lebanese could also help it stand out from the crowd.

In the winery, Gaston has oak trials planned, to assess the exact amount of ageing in barrel best for each vintage, and each variety. 'We don't use additives or industrialized products, so we have to find other ways to create the right flavour balance for each vintage.' While the oak regime has been set to formula until now, Gaston sees good potential in giving it a gentle tweak here and there.

He is also doing some work on bottle reconditioning. 'We did a lot of tests and trials on this and came up with a technique. It is a very

acteristically wary about this, insisting that these start out as limited edition wines only. 'If it is one grape and it doesn't perform in the way we expect that year, then we don't have the other grapes to compensate.' But he agrees that this is an intriguing avenue for the future. 'We just need to have more grape parcels, a variety of sources, so that even though it is one grape on its own, there can still be complexity.'

Where grape varieties won't be singled out, however, is in the white wines. Chateau Musar 'Blanc' is proving more and more popular these days and new vineyards are being sought to meet the demand. Gaston is keen not to under-

delicate, very sensitive process, and only Tarek can do it.' Over the years, as the wines gently 'breathe' through their corks as they sleep and mature, a little – in many cases an almost imperceptibly tiny amount – is lost to the atmosphere. You can tell this from the heady vinous aroma of the cellar, but often enough nothing appears lost at all from the bottle. Sometimes, though, the wines become 'ullaged', the level of wine dropping to the shoulder level. This is not something customers like to find when they purchase their Musar, so the reconditioning process is a gentle topping up of the bottle with the same wine and the addition of a new cork, thus giving 40-year old wines the proper closure to age a further 40 years.

'We decided six or seven years ago that all the red vintages older than 1996 would be reconditioned. It helps a lot – our customers like it because many of them had difficulties extracting the corks on such old vintages. For us it is good because it guarantees their quality.

Marc: taking Musar forward

The changes Marc has made during the 10 years since his arrival at Musar have been dramatic. They pave the way clearly towards Musar's centenary year (2030) and beyond. But ask him about the future of his wine and, with more than a glimmer of his father, he takes a philosophical turn. He is acutely aware that Chateau Musar changes peoples lives.

'In 2013 I had a customer who told me his story. He was studying to be a nurse, and was in his third year at college. One night, he tried a glass of Musar, a red from the 1995 vintage. It was to be his epiphany. The next day, he quit his nursing studies and decided to become a sommelier. A few years later, he was one of six people who won a scholarship to travel to Burgundy. At Louis Jadot, after three days of study, the *maître de chais* set up a blind tasting of four wines and asked the group which one was not a burgundy. The others were lost, unable to spot an outlier, but he recognized it instantly: it was the 1994 Chateau Musar.'

Marc feels that he has a duty, not just to his family and the business but towards the thousands of Musar drinkers who have experienced something really special in the wines. 'My commitment is to make sure that this kind of revelation will still be available to others in the future.'

His greatest passion for the future of Musar is that people understand it as a living product, not a static one. 'How do you define the living organisms on earth? They are the things that change, that evolve, that are dynamic. Anything that is not dynamic and does not change is not alive, it is actually dead. The same applies to wine versus soda, which is static: it is not a living organism. Some wines are produced to be as static as possible in order to endlessly replicate a specific taste: these are what I call manufactured wines and they serve an audience that is specifically seeking replication of this taste. In my opinion, they do not represent what is beautiful or true about wine – or true about life. True wines will be different every vintage; they vary just as the weather does and just as our lives do. They are *alive* and therefore, by definition, will change and evolve: sometimes their pathways are as unpredictable as our own. These are the wines that Musar is seeking and they will take you on a journey as they mature and open up in your glass. They will have the ability to "move" you and to connect with you, one living being talking to another.'

Marc's hope is for an understanding of 'living wine' – not judged by numbers but by its own innate character.

As he was learning the business, Marc spent several years globe-trotting with Serge as he introduced the world to Musar. Back-to-back weeks in Asia were followed by a tour of the Americas, a return to Lebanon (exhausted), then off again. Serge's on-tour tastings were legendary. Far from the polite industry quadrilles of Vintners' Hall in London, they were 'Coltrane and Miles, Ellington and Ella', greeting the audience with constant banter. Never dull.

Serge would nudge and teach by example until Marc eventually developed his own style – recognizably Hochar, but distinctly Marc. He freely admits that he would never get away with the controversial antics adopted by his father, so he developed his own way of relaying the message – without the 'Serge-isms' but with well-delivered (and well-received) clarity.

Marc will say: 'Younger drinkers look for tastes that are closer to what they already know, ie fruit juice. They will prefer wines that still have a vivid memory of fruit. I am not interested in solely tasting fruit in a wine. I want a wine that transports me somewhere, reminds me of things, makes me relive emotions and memories. The more complexity there is, the more filled it is with different aromas and flavours, the better and richer my experience will be. More experienced drinkers will often look for tastes that are more evolved, away from pure fruit and more into all the other realms of taste that often come with older vintages.'

Marc's message, going forward, is that it is mostly senseless to put a rating on his wine as it is a living, changing product. You cannot dissociate the appreciation of wine from time and space: they are linked and intertwined. So if a taster or wine judge gives a wine a rating, it is merely a snapshot of what they think at a certain point in time and at a certain point in a wine's life. He says 'A rating reduces a wine to a number; could you judge a life by a number?'

Don't judge it in a moment, but do enjoy it in a moment. 'In Lebanon, the sense of time is exacerbated. We don't take decisions for tomorrow, we let things simmer until they are ripe, for years and years. And yet, living in Lebanon is also about living every single moment to the full: during the war, we had no clue if we would survive another day. Two hours away – or even 10 metres – there could be shelling. It is the same today: we are safe here until we are not safe. So we have to enjoy each day as if it could be our last. This is the contradiction in how time is perceived in our part of the world. You have to enjoy every single moment despite having to wait such a long time to enjoy it.'

While Marc is happy to embrace the mystical side of Musar, his more pragmatic instinct is to keep a constant eye on the figures. Chateau Musar will always be known for the beauty of its older vintages, so it's vital that a stock of each is kept in the cellars for important events and anniversaries. Maturing the wine for seven years – even in the Hochars' expansive storage bays cut into the bedrock at Ghazir – will eventually take its toll on space as not every vintage can be kept in bulk indefinitely. But the truth is, increasing sales are beginning to deplete the stocks accumulated during the civil war years and up to the mid-2000s. 'Consumers will have to get used to buying our wines in their first year of release and age them in their own cellars,' says Marc, becoming steely: 'They will no longer easily find specific older vintages. Previously, the price of our wines depended solely on the age of the wine, no matter how much we had in stock. Today, we increase our prices as the level of stock becomes low, and release our older vintages in much smaller quantities so we

For Ralph, wine is an ambassador for his country – with his wines comes a consciousness of Lebanon, too.

can keep some for the future and not run out. This has been difficult for the market to understand as some younger vintages have become more expensive than older ones, but for us, and at this point in time, pricing according to rarity makes much more sense.'

These and the many other changes Marc has helped in instigating – new brands like 'Levantine', expansion onto wider global markets, better teamwork and data gathering processes – have transformed the business dramatically since 2010 and helped create a great looking future for the fourth generation of Hochars.

Ralph: maintaining Serge's legacy

There's a lot of the Phoenician in Ralph. He has his uncle's passion for travel and for meeting new people. Communicating the love he has for his family's wines is his motivation – whether by Twitter or by taxi to Tianjin, reaching out is the way he finds new customers, new feedback and new ideas. 'It's such a great way to learn about how we are perceived. And who we should tell our story to next. We can't do this by staying at home!'

Ralph remembers how he felt as an 11 year old at the winery in Ghazir, the clinking sounds of the machines, the way the bottles filled with wine, the snug feel of the cellar with its sleeping oak barrels. He remembers listening to Serge talk. 'He'd say to me: "I'll give you a mission. I don't care how you do it, but you have to do it." So I'd have a task and have to work out how to get it done. It feels like I'm running those same missions now.

'My uncle did not like the idea of selling, but he wanted to find out the best way for an understanding of our wine to be spread. He found that to build up a close relationship with our customers, so they became friends, and he could explain each vintage and teach them his approach to winemaking, worked well.'

For Ralph, Serge is his guiding light. It is as true for him as it is for the rest of the family that the devastation felt at losing Serge is what drives him to keep his memory alive. 'I want to show our wines in the same way Serge did, to continue his legacy.'

When Serge died in 2014, he had a full programme of events booked for the year following his visit to Mexico. The family was in shock. To drown in the wild sea on New Year's Eve, the last day of the year, at the other end of the world, presented them not only with crippling grief but the most difficult logistics imaginable.

As Ralph explained: 'It was very, very hard for all of us, especially for my father. We were all very close and so full of sadness. And Serge had such a full programme planned that we did not know what we could do about it. But we decided that we would honour everything that he arranged. Everything.

'It was hard work that year, 2015. It would have been hard enough for a 35 year old, but Serge would have been 75! We each took on some of his tours. We travelled to Europe, the UK, to Canada. My father even held a tasting at Cambridge University that Serge had planned.'

Ralph was close to his uncle and it was this time that really brought to him a sense of the

Ralph, Ronald, Serge and Gaston flanked by their legacy: 1,000 and more bottles of vintage Chateau Musar.

importance of his task. 'We had to keep going with everything he established. We had to keep up the momentum. In terms of our family, the culture of Musar, nothing had to change – and yet we had to keep reaching out to the future.

'Serge had his own kind of "social media". He developed his own way of sending out his message, and he had many followers. Today, we don't have to travel as much as he did; we have apps to tell everyone what we are doing. But we travel anyway, we must, as this way we reach out to customers properly. Serge always believed in Lebanon and never saw himself being anywhere else. It is a very hospitable country, but not one that enough people come to; we have to take the message out ourselves.'

Times are unsettled once again in today's Lebanon. And this is a great worry for Ralph and his father Ronald. The economic affects of more in-fighting in the country are likely to be dramatic. Internally, the banks have frozen funds and the roads become blockaded again. Commercial movement is thwarted and purchasing power is stretched thin. Externally, wealthier countries (the US, for example) begin to doubt their investments in Lebanon. But, as Ralph says, wine is a vector for peace. 'It is universal, it is a good ambassador for us. Aside from the revolution in Lebanon it is a country of peace. Despite the things that happen (and probably always will) it is a very welcoming place. And wine does not come from people who try to create tension. It is about tolerance. It will be a struggle for us all for a few years, but at the end of the day, people will still enjoy wine: they will still eat and they will still drink.'

Around 85% of Chateau Musar's wines are sold on the export market, so these sales should be stable during times of political turbulence. Ralph believes that Serge would have

been proud of this balance in 2020. 'He'd be happy with our sales figures which have had a good increase in the last five years. And it is good. We are present in more and more countries around the world.'

There are still challenges, and new countries – some of Europe, southern India – where Musar is not yet known. But these are all ready to be met and explored in the near future.

Jane Sowter: the whiter side of Musar

Jane is as much part of Musar's future as she has been its past. She is a strong believer in the Chateau's white wine, Musar Blanc. 'From the very first time I met Serge, he was always on a mission with this wine. From the first time I met him, until the last.' Desirée Khoury, a stalwart of Musar's Beirut office who has devoted her working life to developing the distribution business in Lebanon, feels an equal loyalty to the white 'underdog'. For her 'it is always the wine I go to for an important celebration in the family, it just has such a majestic quality to it'.

'Not many people understood at first, but he wanted them to listen,' continues Jane. 'They were seduced by the red. They got engaged with the red; they got married with the red; they wrote their life stories with the red, but it took some time and explanation to understand our whites, how best to serve them, and with which foods.'

Musar Blanc is every bit as age-worthy; the layers of nutty, honeyed complexity it develops as it is cellared are legendary. And today's sommeliers are loving its remarkable ability to match food. (Serge wasn't a fan of food and wine pairing, but his white is a fearless partner for Lebanon's trickiest food combatant, 'tabbouleh' – a mix of mint, parsley, bulgar wheat and zingy lemon – so he perhaps should have

'My man in the UK', Jane Sowter has been involved with Musar's sales in the UK, Europe and Canada since 1988.

been more proud.) Its greatest 'disadvantage', apart from going against every expected wine convention and being best served after the reds, has been that it is made from grapes not generally known. But today the world is hungry to learn about quirkier local grape varieties, and Obaideh and Merwah are beginning to be welcomed as they deserve to be.

'Serge believed in his white wine, with the depth of a thousand years of culture behind it straight from the Phoenicians,' says Jane. '"This is my latest red," he would exclaim: "It is way bigger than my reds. The only thing that it is white about it is the colour!" I think he would be happy now that the white wine is an important part of our portfolio and sold all over the world to Musar fans and collectors.'

For the first time, with the 2010 vintage, Chateau Musar white has been released in magnum. 'Serge would have been overjoyed to

The incredible potential for growing luscious fruit in the Beka'a Valley makes it hard for the Hochars to imagine growing grapes anywhere else, but occasionally, they do...

see it in this format,' says Jane: 'Magnums mean that a wine is being taken seriously. This wine ages beautifully for decades, and magnums reward you with even more time.'

Musar at home and abroad...

With torrid times of revolution and civil war not always behind them, the cousins (Gaston, Marc and Ralph) could be forgiven for looking abroad for new Musar vineyard destinations. It was a pet project for Serge too. He and Ronald explored the idea of opening up in California in the 1970s, where the climate is similar and the threat of earthquake considerably less disturbing than Lebanon's war. But it never happened.

'In the 1980s,' says Gaston, 'we approached a domaine in Burgundy. But that also didn't happen.' Today, the leaning is more towards potentially buying a vineyard in Mediterranean Europe – Spain, Italy or Southern France – to countries with a similar climate but nearer to home, because ultimately the family will always come back to Lebanon. 'It's just an idea at the moment, not serious,' says Gaston: 'We have too many other issues here to worry about.'

Another significant concern for all the team is who will follow in the winemaking footsteps of Gaston and Tarek. Tarek, like Gaston, is in his early 50s; for him, Musar's future still holds excitement and possibility. As well as maintaining the legacy of the *grand vin* there is the chance to capture the true beauty of Lebanon's *terroir* in wines that exhibit the purest and most authentic expression of each grape variety. For Tarek, younger, more vivid expressions of Beka'a fruit are what's important, and these will also reflect the benefits of Musar's organic

All cool for the future – Marc and Ralph in the shade of the old winery (below) and the snow still present on Mount Lebanon in late June.

viticulture. But it is early days, and he is not at all ready to hand over the winemaking reins: 'To work with these wines you must be prepared to learn over time, as I did with Serge. What we do here does not come from the obvious choices taught in wine school. It takes sensitivity and very much care.'

There is still time to decide: the fourth generation is not yet waiting in the wings. Gaston's children (Serge 27, Isabelle 24 and Camille 21) are busy with jobs in technology, pharmacology, or still studying. Marc's daughter Mia is only 14, and Ralph's children Natalia and Matteo are 13 and 9, so for them it is too early to tell. With luck, one of these children will be a small version of Serge, with that same passion for wine that will keep the Hochar flame burning.

Whatever happens, it is certain that Musar's story will continue. Its fabulous wines will lie maturing until some are 100 years old, and then, what a day. A day Serge would have loved. The day to taste a century-old Chateau Musar, perhaps in 2059, sip, by sip, by sip. It will have even more of a story to tell. 🏵

Chapter 10

TIMELINE OF VINTAGES

Tasting notes year by year

SERGE-ISM

For me, 'identity' is a key word. Some people play poker and so do I, but the biggest gamble is to make wine. If you don't want to gamble, you have to kill it. Stop it from being alive. In my wines there is a life you cannot stop. If you cannot stop it, it will live its own life. Fining removes those substances you strip if you want your wine to have less of a life. The same is true with filtering – it removes some of the living organisms and reduces the potential for organic chemistry. These might be the nutrients that seduce your brain, that bring nuances to the wine. If each grape has 100 different dimensions, imagine how this is compounded in a wine! This is why Chateau Musar wines are so addictive. The different dimensions communicate with your brain. The complexity of these dimensions increases during the two years each grape variety spends by itself before meeting other grapes.

SERGE HOCHAR

MANY FACTORS COME into play when tasting Chateau Musar vintages. Was it a warmer Cabernet year (1999, 2000, 2005)? Was it cooler, making Cinsault more influential in the blend as in 1998? Or was it a Carignan year like 2003? Or will grape prominence be less relevant than the dodging of bullets across the Beka'a in a vintage we're lucky to see the appearance of at all?

Through the tasting notes that follow, there are notes on the vintage, the weather conditions and the family events that helped shape the wines. They also chart the mindset of the winemaker, Serge Hochar, as he became less influenced by his background in Bordeaux and increasingly confident in the wines emerging from his own *terroir* in Lebanon; how he became frustrated with the politics that prevented him from accessing his vineyards, then relieved as the freedom to harvest on time returned.

Hochar family wines live for a long time and their flavours linger in the mouth and memory long after the bottle is drained. Some of these wines have been tasted *à point*, at the peak of perfection, but many have yet to reach this moment, and they are judged on their potential. Even more fascinating, perhaps, are the notes on wines that are perhaps just past their prime, but still delivering fascinating nuances that astound the taster – some of these wines elicit the most evocative tasting notes of all.

For the Hochar family, a new wine, however shapely or pleasing, is always pregnant with

THE TASTING PANEL

Jancis Robinson MW (JR) Michael Broadbent (MB) Bartholomew Broadbent (BB) Steven Spurrier (SS)

Additional notes by Serge Hochar (SH), Gaston Hochar (GH), Marc Hochar (MH), Tarek Sakr (TS) and Susan Keevil (SK).

promise for the future. Many of the notes below show the best that the wines can be – but they also underline the fact that for each taster Musar's most sublime moment will be different.

Although the Hochar family does not believe in the concept of defining a wine with numbers, ratings have been included in here to give some guidance. Marc is keen to point out that these figures, although helpful, are sometimes very different from the feedback received from professionals at tastings in the US, Europe or Asia. 'But that should not come as a surprise,' he says: 'Tasting is always a very personal matter.'

Jancis Robinson MW, Bartholomew Broadbent and Steven Spurrier award marks out of 20 for each wine. Michael Broadbent prefers a five-star scoring system. All notes were taken in 2018/19 with the exception of Michael Broadbent's. Jancis' notes are reprinted with the kind permission of JancisRobinson.com.

CHATEAU MUSAR – RED

One third each Cabernet Sauvignon, Cinsault and Carignan, aged for a year in French Nevers oak.

2013 Dense, deep red, still hints of purple at the core and little age on the rim. Marvellously warm Mediterranean fruits on the nose, autumnal in style but youthful in their rich, robust, vigourous presence. The essence of Lebanon in a glass, rich with spice and cooling breezes, the dry, herby fruit flows with energy over the palate before giving way to a finish that confirms the structure, fruit and tannins blending exuberantly with intriguing hints of much more to come in the next decade. 2020–40. **18** (SS)

2012 Dense red, still very young, with thick 'legs' on the glass showing richness to come. The nose is still quite closed, with autumnal black stone fruits that gain exuberance and spice on the palate, plus an almost sweet warmth. Needs to lose 'puppy fat' and move into its second decade before showing its character to the full. 2022–35. **17.5** (SS)

2011 Fairly deep ruby. A denser, more concentrated, more linear nose than the 2010 served

CHATEAU MUSAR **169**

alongside. A certain dustiness of aroma. This seems more obviously Cabernet than most Musar reds. Still quite youthful. Falls away a little on the end. Well judged but less distinctive and less obviously Lebanese than some vintages. Decent balance. Certainly not over the top. 2020–34. **17** (JR)

2010 Bright ruby. Heady, rich and gamey on the nose. Very hedonistic and accessible with real density and lifted slight gaminess. An attractive singed character on top of very ripe fruit. Great balance and freshness (but not excessive acidity) on the finish. Very neat on the end. Something rather wild and Romany about this. Already accessible even if I'm sure it will develop even more complex tertiary aromas. 2018–28. **18** (JR)

2009 Mid-crimson. Light, sweet start. For the moment it's a little simple. Bright, ripe red fruit. Slightly burnt edge. Lots of tannins. Quite dry and youthful on the finish. Much longer bottle age needed. By far the most youthful wine in this vertical tasting. 2028–50. **17+++** (JR)

2008 *'After blending, Tarek was raving about this vintage. I asked him to add 0.5% of any wine he liked and we'd taste it again in two weeks. An act of faith or* la folie? *We tasted it again. He raved again; the wine had changed colour and bouquet.'* (SH)

Heady and beguiling with a gamey note on the nose. Very sweet and young; well balanced on the palate. No hurry at all to drink this youthful ferment. Very tight, dry wine that is so much younger than any other vintage tasted. SO much more youthful than a Bordeaux 2008! 2025–45. **17.5+++** (JR)

2007 *'This was a good year for Cinsault. It took me 20 years to fall in love with this grape. People generally think it's a cow, an over-producer. But restrict her yield and it's another matter...'* (TS)

Mid-garnet. Pale rim. Paler than the 2006. Light, tea-leaf nose. Sweet start. Dry and powdery finish. Very unformed. I could easily enjoy this with Greek/Lebanese mezze – in fact, these flavours strongly remind me of dolmades – but ideally with all these fine tannins I would continue to cellar this vintage. Quite long. 2025–42. **17.5** (JR)

2006 *Cold winter (two weeks of snow) and cold summer – just 10 days of normal summer temperatures. Musar wines are usually released seven years after harvest but in 2013 the 2006 was still far from ready so it was held back until spring 2017. (SK)*

Another new chapter in terms of density of colour. For the first time (tasting from old to young) this could be described as crimson. Seriously heady! Smells exotic and mature on the nose but this doesn't transfer to the palate. Compared with the older vintages I have just tasted this of course is a baby... Masses of concentration and allure but, although it would give pleasure, it would clearly be sensible to keep it for quite a few more years. Lovely combination of fruit and freshness. Really vivacious. Sweet and rigorous. 2020–40. **18+** (JR)

2005 *'A Cabernet vintage – not pretentious or aggressive.* Une profondeur interessante. *It is powerful, but this wine is still young, a baby.'* (SH)

Full garnet-red colour, mature. Red and black autumn berries on the nose, smooth and ripe with an appealing warmth, cinnamon and figs; the spicy fruit from the Carignan grape adds a savoury smoothness to the firmer Cabernet. Tannins are resolved, mature, while the Musar acidity keeps the finish fresh. A fine, open wine, seemingly at its peak now, but one can never really know with Musar. To 2025. **17.25** (SS)

2004 *'At the start of my second decade at Musar I had already learnt a lot and was beginning to understand you never know enough!'* (GH)

Fine, vibrant, carmine red, very youthful looking, the freshness of the Cinsault grape showing on the nose. More red than black fruits, leading to a mouth-watering, even succulent fruit on the palate. Beautifully lively at 15 years old with an exuberant elegance that is so attractive that one almost forgets the wine's innate seriousness, which is impossible to miss on the perfect, lifted harmony of the finish. To 2030. **18** (SS)

2003 *'I think of the 2003 red as a great vintage for Carignan.'* (GH)

A vintage with high alcohol and acidity. Rainiest winter in 15 years, then not a drop of rain from April throughout a hot, sunny summer. Heatwave during flowering reduced the harvest by 30%. (TS)

Mid-garnet. Quite a bit deeper and younger-looking than even the 2001. A little marine influence (oyster shells?) on the somewhat reticent nose. Rather different from many of these Musars. Sweet start and very vigorous and fun. If I had not tasted vintages back to 1961, I might have taken this for a fully mature wine. It already gives masses of pleasure with some milk-chocolate notes and quite a bit of light dry tannin on the end. Excellent drive. Long and dramatically big, but with good freshness on the finish. 2016–35. **18.5** (JR)

2002 *'This vintage is now tasting more like a Bordeaux blend than I had remembered. Great tannins and structure.'* (GH)

Just a hint of brick red on rim showing age, otherwise a light, glowing ruby colour. Rather mature sweet nose that fills the glass and room, very autumnal; treacle and brown sugar are joined by Golden Syrup and molasses aromas. Beautiful mouthwatering acidity; a touch of banana; lovely spicy, hot mouth feel. Great length. The superb acidity will give it a very long life. Very full flavoured, yet elegant and distinctly Burgundian. **17** (BB)

2001 *Cold and rainy spring, but a warm and very dry summer. July and August were hotter than usual and picking began unusually early on September 3rd. Tannins and acidity were low, and the thick-skinned Cabernet grapes fared best. (TS)*

Cork was unwilling to be extracted so we had to taste from the bottle in which it floated. Mid-rust red. Pale rim. Rather heavy nose. Dense and mushroomy. Quite a contrast to the ethereal 1999. Dusty chewiness on the end after a slightly obvious sweetness at the start. Lacks a little bit of freshness, but there is some interesting maturation. 2013–28. **17-** (JR)

2000 *'A Cabernet Sauvignon year. One of my clients called it "a lioness" of a wine.'* (SH)

Reasonably mature. Brick-red rim, autumnal faded red bowl. Mouthwatering acidity on the nose that can be smelt a foot away. Lovely aromas of damp autumn forests, decaying leaves and wood combine with a slightly chocolatey sweetness. A little spirit developed over the course of the evening: almost a Drambuie or Grand Marnier character. Slightly sweet on the palate with very long, classic autumnal Musar notes, soft, mature and Old World, braced by that lovely acidity. Still full flavoured and crisp on its long, lingering aftertaste, finishing with notes of liquorice and sherbet. **18** (BB)

1999 *'A Cabernet vintage. One of the greatest years of the century. Masculine.' (SH)*

This is the basic palette from which Serge Hochar, no dry academician, creates his blends: first, a Cinsault from a single, soil-rich vineyard in Aana: deep core, young Provençal scent; delicious flavour, wonderful richness and flesh. Soft tannins. Next, a Cinsault from Ammiq's very gravelly soil: fragrant, lighter style, more charm. Then Carignan from a lighter soil in the Aana district: deep, velvety; spicy garrigue scent (I was reminded of Corbières in the Languedoc), flavour more aromatic. Lastly, Cabernet Sauvignon from Kefraya's rocky soil: very deep purple; sweet, lovely crisp flavour, excellent tannins and acidity. End taste of violets and blackcurrants. Tasted in the Chateau Musar cellars, December 1999.' (MB)

Warm autumnal red colour; mature yet clean and fresh, warm spices on the nose. Seems more 'southern' than the 2000, yet retains the classic Musar acidity, all of a piece, totally open now but will keep evolving to 2026. **17.75** SS

1998 *'Tarek said: "Let's bottle at the end of the second year. Let's try!" We did 20,000 bottles like this, then the rest a year later, as usual. We tasted the wines four years later. Something was missing in the earlier-bottled wines. They had an obvious Chateau Musar character but were not necessarily better. Identity is permanently important. We went back to the original formula.' (SH)*

Cool year, Cinsault dominated. Cold and rainy until June, with a sunny, dry summer. Harvest began on September 9th. (TS)

Pale, jewel-bright garnet. Light, spicy, particularly well-integrated nose, already well developed. A sweet start on the palate; the opposite of heavy; a really lovely wine: fresh, sweet, with some very slight mintiness and a dry finish. Very long, and nicely mature: seems just right now. It will go beautifully with food. One of my favourite wines in this collection. Is it the Cinsault I like so much? 2012–30. **18.5** (JR)

1997 *'We had two harvests – before and after a week of heavy rain – and made a big, big wine. Two years after bottling it changed direction and began to create heavy sediments. It was the biggest surprise in my life. This wine was behaving like a human going through an adolescent crisis. Which way is it heading now? The only thing I know is that I trust my wines. It is already an interesting wine, and starting to taste beautiful.' (SH)*

Very light in colour with maturity on the rim: salmon ruby. Surprisingly intense and concentrated nose with strong ash and autumnal smells; great acidity, damp leaves and old bonfires. On the palate, very sweet fruit and a long finish that goes through several evolutions (one developing old pomegranate flavours) in the longest aftertaste of any Musar. Full mouthfeel defies the lightness of the wine. Almost chewy in texture. Intense volatile acidity. **18** (BB)

1996 *'The experiences of the previous harvest pushed us to protect the wine with a little bit of sulphur during racking. At the time I felt the 1996 was too weak, and that we should release it early. It has grown to become a wine of poise, beauty and elegance.' (SH)*

Fine, mature garnet-red, quite diffuse at the rim but no browning. Lovely florality on the nose for a wine of this age, with lifted freshness thanks to the presence of the Cinsault grape in the blend. Fine clarity of red fruits on the palate that hint of claret, and a natural warmth that looks to the Southern Rhône. But above all this is Musar: gamey, freshly leathered, ready in its half-bottle state for several more appearances on stage. To 2026. **18** (SS)

1995 *'A great year, great weather. Mid-summer heat with lots of sunshine and cloudy evenings. Perfect. The grapes came in at full maturity but some did not finish their fermentations; these were racked (pumped into different vats) but not sulphured – this was because of my No Touch Philosophy. It took some grapes one year to ferment! The volatile acidity went up, of course. We bottled it in 1998, but when we wanted to release it in 2002 it was too acetic, not ready. So we released it alongside the 1996. Now, people tell me it is my greatest vintage.' (SH)*

Mid to deep red, tawny rim. Good for its age. Attractive autumnal berry fruits on the nose and natural warmth still present with lifted, elegantly textured fruit and good length on the palate. Still firm, with a clarity of expression that will remain for another few years. To 2025. **17.5** (SS)

1994 *'Hot summer in the Beka'a Valley: 40°C in August and September. One can smell the singed grapes in the wine; remarkably sweet, lovely fruit. Two recent notes, identical. Last tasted August 2000.' (MB)*

'A vintage that was great when it was released in the early 2000s. One US importer loved it so much that he decided to buy all our stock. A couple of years later, it went through a difficult phase. This is typical of Musar. We explained this and told the importer that all he had to do was wait a little. But he got upset so we agreed to take back some of the cases. Ten years later, we met by coincidence a day after hosting a tasting with the '94, which had been voted the best wine of the night. The importer could not believe it. He wanted to buy his cases back, but the price and availability were now completely different...' (MH)

Mid ruby; pale rim. Heady, animal nose. Jewel-sweet fruit and quite an aggressive attack; still quite edgy on the palate. At the moment it could do with a little more concentration in the middle, but it's certainly sweet at the beginning. Not the most refreshing but very confident and vigorous. Reminds me a bit of the 1980. Some light tannins evident on the end. Still resolving itself but admirably transparent. 2012–18. **17.5** (JR)

1993 *'A classic vintage for the reds.' (SH)*

Red-mahogany, diffuse rim. Autumn red fruits quite pronounced on the nose; good warm depth on the palate. More natural richness but less clarity of expression than the 1995, showing Rhône-ish flavours (the 1995 has more claret), but has impressive robustness still and good length. To 2023. **17.25 (**SS)

1992 *[No wine available.] 'The weather was terrible. We had a winter that lasted until June. Hail came. Biblical weather! I tried to make a Chateau Musar, but two years later found it wasn't up to standard, so we distilled it. The best grape that year was Obaideh.' (SH)*

1991 *'A classical year, with Cinsault predominant, which made a very interesting wine. We had good vintages throughout most of the 1990s, after all those difficult years. With this peace, I settled down my mind. Tania came home. Gaston got married here at the winery. We had 400 people and served the 1966, the year of his birth.' (SH)*

Fine mahogany/deep tawny red, slightly burnt looking but clear to the rim. Very striking red fruits on the nose showing both richness and an astounding freshness for 28 years. Lovely natural warmth and sweetness of fruit on the palate, the touch of sweetness enhancing the richness of texture, while a line of acidity and tannin firms up the finish, still has energy and depth. To 2025. **17.75** (SS)

1990 *'A mild winter with very little rain and less than average snow fall on the mountains was followed by one of the coolest summers on record. After an early flowering we expected an early ripening, which did not happen. We had to bring the grapes in anyway as the war flared up again and the roads were blockaded on September 28th. (SH)*

Mahogany-red, tawny rim, has clarity but definite maturity; bottle at mid-shoulder level. Attractively concentrated red fruits, dry roses and even a rose-hip sweetness on the nose which continues expressively onto the palate giving rich/dry flavours due to inherent acidity which adds length. Perhaps just passing its best, while the sense of vineyard is still very much there. **17.5** (SS)

1989 Tawny-red with a pale and watery rim. Still some vigour and richness on the nose but rather 'high tone' with volatile acidity/acetone evident. A bit softer on the palate, but gives the impression of a very hot year difficult to control in the cellar; might have been impressive in its youth. **15** (SS)

1989 Carignan *'Carignan was so impressive this year. Its transition from juice into wine was incredible. I decided to produce a special bottling of 100% Carignan for my 50th birthday. It is not Chateau Musar but it is very, very fascinating.' (SH)*

Good colour. Sweet, full, rich and uplifting. (MB)

Mid ruby; still good depth of colour. Particularly cheesy nose – very different from the usual Musar. A tiny bit tart, rustic, lighter than most but still very appealing. Could do with a little more fruit and purity but I like the maturity. 2000–22. (JR)

1988 *'Great weather that year, and the wine is developing quite well. Michael Broadbent chose it for his son Bartholomew's wedding, and I did for my son Marc's wedding. Life in Lebanon was not so hectic. A normal year, politically speaking.' (SH)*

Tasted at *Decanter* magazine's Millennium Dinner: very burgundian, perfect with duck. Sweet, soft and, yet again, delicious. An identical note made at my 75th birthday dinner at Vintners' Hall, London in May 2002 and three times since. The most perfectly mature red imaginable: *à point*. I am busily drinking it at its peak.' **Five stars.** (MB)

1987 *'A simple year. Still ageing, so who knows?' (SH)*

1986 *'For the reds, a Carignan and Cinsault year. When the time came to bottle it I said: "No, stop! The wine is too perfect, and I hate perfection. So we added 1% of a vinegary vin de presse. Today, the 1986 red is the best I have ever produced.' (SH)*

Orange-tinged; sweet, chocolatey nose opens up richly; full, ripe, Pomerol syle, with quite a bite. (MB)

Mahogany-red with fading, tawny rim, plainly mature but still with an impression of youth for its age. Warm, very southern nose, like an old-fashioned Châteauneuf-du-Pape with a line of acidity that this would lack. Rich, satisfying and very Rhône-like on the palate, still with vigour and bitter chocolate. A lovely wine, now at its peak and beginning to slowly fade. To 2024. **17.25** (SS)

1985 *'A big wine like my 1964.' (SH)*

Mostly Cinsault, picked early, long fermentation and no wood used. Lean and lively, with good acidity. Then 12 years later (in 1999) its nose reminded me of a ripe Château Brane-Cantenac. Very tannic. (MB)

1984 *'In April I met The Wall Street Journal correspondent who asked me to predict the year. I said "my nose predicts no year". Six months later I understood why. We could not harvest the grapes in time due to the war. By the time they'd finshed their long journeys to the winery, the grapes were so over-ripe that they were already hot and fermenting.*

The wine I made from them was like no other wine I have made before or after. I bottled it for the hell of it, and I trusted the 1984 to become whatever it decided to be. I used to call it my Madeira, then it became like a port, but now it has become a beautiful, elegant wine, a wine of the brain and a wine of the mind.' (SH)

Level mid-shoulder, cork saturated with wine, beginning to crumble. Deep mahogany red with browning rim; a rare Musar looking older than its years. Yet its Mediterranean warmth is still there on the nose with a surprising richness that seems attractively 'fortified'. The same extra concentration on the palate, finishing rather like a dry Banyuls with the typical Musar grip of acidity. A great wine to be savoured with dry cheeses. To 2024. **17** (SS)

1983 *'The vintage for which I fell in love with our wines. I started drinking this in my mid-20s.' (MH)*

Full mahogany-red, still vibrant and looking good for its age. Rich red fruits on the nose with lots of vigour and warmth, not heavy, just the richness of a Châteauneuf-du-Pape with perhaps an extra layer of fruit from a hot year. The Musar acidity is there to refresh the finish, still extraordinarily vigorous for its 36 years. A richly satisfying wine. **17.25** (SS)

1982 (Low shoulder.) Deep mahogany-red fading rim. Quite rich, with attractively fading red fruits on the nose, which still has some vigour and is more elegantly expressed than the 1983. Less rich on the palate but with more length both in texture and flavour, perhaps a touch of volatile acidity now, but would have been a superb wine in its youth. Cannot improve but it has kept its memories. **17.5** (SS)

1981 *'A fantastic vintage. One of Serge's favourites. (MH)*

'A year for Cabernet Sauvignon. More opulent, more brilliant than the other cépages *that year.' (SH)*

(Level: high shoulder.) Fine, deep mahogany-red, tawny rim, still vigorous and not fading at all. Very good impression of warm fruit on the nose, quite earthy but pure and naturally ripe youthfulness that encourages one to move to the palate which confirms the richness and balsamic warmth. Acidity is there as a back-up but not taking over. This is a really fine wine, quite remarkable from a half bottle at 38 years old; the sweetness of fruit will remain for another few years. **18** (SS)

1980 *'A delicate vintage, developing much more character than we would have first thought.' (SH)*

Marked bottle deposit on the shoulder. Pale rust-red. Hint of decay on the nose which may well be cork-related (although the cork itself smells clean). Very vigorous, palate-grabbing start. This is full of life with light tannins on the end. Very impressive in many ways even if it's still quite youthful. Really very difficult to assign a score to. I'd love to taste another bottle of this to assess whether that slight note of decay is because of the wine or the cork. Wonderfully impressive in every respect other than this (which I will ignore in choosing my score) and tastes as though it has a great future. The note of decay dissipated. 2008–30. **18+** (JR)

1979 *'It is an easy-going vintage: classical and happy to be so. Ageing has been consistent. I'm deaf until I open a bottle of this, because each wine has its music. This wine would not be Wagner, but maybe Schumann.' (SH)*

...misleadingly pale, fully mature; nose like squashed strawberries; sweet, long, delicious. (MB)

Vibrant red with a hint of mahogany but still red on the rim. Astoundingly young and vigourous with warm red summer fruits and a definite hint of roses on the nose; has the feel of a wine at least 20 years its junior. The sweetness of fruit continues on the palate with a backbone of tannin to firm it up – without losing the old-fashioned 'burgundian' richness and just the right level of acidity to refresh the palate for another glass. This must always have been an exceptional wine, as it remains today. **18.5** (JR)

1978 *'A great vintage. I thought it would be special, if quite awkward, but it became great.'* (SH)

Very mature orange brick-red notes. On the nose, slightly sweet but beautiful, with lovely aromas of old cellars, leather armchairs, old bibles and very old pews in an ancient church – a delight to connoisseurs of very old wine! High acidity should carry it for years. Palate very thick and syrupy in texture. Enormous flavours. Spicy. A little tannin showing on the long finish with perhaps a touch of tartness. Completely mature – over-mature to the uninitiated but lovely to the diehard Musar fans. Perhaps a taste of copper coins on the finish. **17/20** (BB)

1977 *'An important year because I settled on the Chateau Musar recipe: more or less equal quantities of Carignan, Cinsault and Cabernet Sauvignon.'* (SH)

Sweet, assertive, still tannic (July 1999); singed flavour, long tannic finish (December 1999). Sweet, rich, fairly full-bodied despite misleadingly pale colour (April 2000, four stars). (MB)

Mahogany-red, tawny rim, still with some vigour but fading. Surprising freshness and natural richness on the nose, not the depth of 1981 nor 1979 but Mediterranean warmth is plainly there in a slightly leaner form. What other half bottle today could show such warmth, clarity and sense of place after 42 years? **17.5** (SS)

1976 No wine made due to war.

1975 Pretty pale ruby. The first wine to smell perhaps a little too old. Not as fresh as some though it delivers lots of naughty pleasure. Dustiness slightly trumps the fruit. It's a tad dried out on the end

though the flavours are a delight. Clean finish. A bit like a mix of mature vintage port, lemon juice and water. I'd like just slightly more fruit concentration in the middle. 1990–2020. **17.5** (JR)

1974 *'My '74 I wasn't happy with. But it developed! In 2000 I tasted it again. Now I think it is nearly ready to drink.'* (SH)

Good depth but still light, with a mature looking, orangey rim. Very vibrant, sweet fruited nose, alive and drawing attention to itself. Very old, elegant fruit on the palate, with a touch of cheese rind. Extraordinary texture: velvety and syrupy at the same time; deep, with high acidity and a very long finish. Biscuity flavours ('Playbox' biscuits: if you are English and of a certain age, you'll remember licking the sugary coat off those). Although past its prime, it is fascinating, like an old man with many stories to tell. You can imagine the war time conditions through which this wine has aged, enduring much and remaining strong, until now as it begins to lose its way. **19** (BB)

1973 No longer available.

1972 *'One of the greatest years. Clive Coates of the Wine Society said our 1970 was very good. But I think the 1972 became better. Was this result of gambling or a guessing game? Who can say?'* (SH)

In 1979, Serge told me that his red wine was made with a blend of Cabernet Sauvignon, Cinsault, Merlot and Carignan. Already fairly mature, ripe, fullish, reminding me of a South African Cape or soft Coonawarra red. Four notes in the late 1970s to mid-1980s, including a rich, earthy bottle at my daughter Emma's confirmation lunch, and in 1987, in the US with Le Tournedos de Bison, of course! A splendid combination. Most recently, attractively coloured, very sweet on the palate, lovely flavour, good length, complete. Last tasted pre-sale at Christie's, April 2000. (MB)

A legendary vintage, still with amazing depth. Fully mature, browning orange rim but obviously still tons of fruit. Massive unique nose. Smells of old wood and sweet fruit. Totally evolved and mature but not really showing signs of ageing. Age, yes, over-maturity not at all. Fascinating, barnyard, old gumboots. Apricot. Very big. Seems higher than the stated 13.5% alcohol. Very long, very intense, full flavour. Silky texture. Very alcoholic finish. Past its prime but still worthy of its reputation and stature. **19.5** (BB)

1971 No longer available.

1970 *'Mistakenly, the most Bordeaux wine I ever made. This was unintentional. I was trying to reflect Lebanon in all its guises, all its aspects. It had more Cabernet Sauvignon than any wine I've ever made, but this was when I could play as I wanted. There were no rules to command me, so I just blended according to my palate.' (SH)*

In 1983, richly coloured; pine needles and vanilla; soft, velvety, well-balanced, yet powerful and plump like Château Petrus. Labelled 'Réserve Rouge. Grand Cru', still deep and rich, complete. **Five stars.** (MB)

Dark, shaded ruby; just the sort of colour I would expect from a 1970 red Bordeaux. Light, very Cabernet nose. If I were given this blind I would be tempted to head straight for Bordeaux, though it would have to be a comely château such as Palmer perhaps, and a fairly ripe vintage. Perhaps 1961? But then there is probably more sweetness and a bit more exoticism than one would find in a Bordeaux of this evident age... Mouth-filling and again that note of violets, candied this time. A little dusting of tannins on the end. Lovely wine. 1985–2022. **18** (JR)

1969 *'Speaking of the red, in those days how many things we knew? It was early stages, only. I remember that we had a good cold winter and a very nice, sunny summer. The grapes reached full*

maturity, so it was an impressive vintage. Amazingly enough when it was young it was over the hill and as it's getting older it is back on the hill. This is why I say my wines get younger every day!' (SH)

First noted at the Bristol Wine Fair in 1979: it was already very mature-looking, there was very little red; extraordinary character like an overripe singed Pinot Noir; dry, surprisingly full-bodied and swingeing tannins. It reminded me of a South African Shiraz. Over 20 years later, orange-tinged; slightly raisiny bouquet; sweet, nice weight. No harsh edges. Delicious. Amazing development in April 2000. **Four stars**. (MB)

Pale rust-red with a very pale rim. Much more obviously sweet and less ethereal than the 1961. This smells like delicious, old-fashioned red burgundy (possibly out of Algeria and the Rhône). There's a bit of alcohol on the end. Thoroughly rumbustious. What an impact! Sweet violet compote. Fine, long-polished tannins have almost completely receded. Lots of fun. Very slight powderiness on the finish. 1984–2024. **18** (JR)

1968 *Never sold. 'We missed this year. I tried again to drink the cuvée we made but we bottled very little and it did not work out very well – I don't remember at all why not. But I do remember that Marc was born in November.' (SH)*

1967 *'What did I do, and why did I do it? I was still under the influence of Bordeaux and started blending wine almost as soon as it was in the barrels. I think I realized then that although the Bordelais make their assemblage in the December after that autumn's vintage, that my wines would need time to develop, time to express themselves.' (SH)*

The enterprising Serge Hochar presented his wines for the first time in England at the Bristol Wine Fair in July 1979 which is where I first met him and tasted his wines: three very dry white and three red.

Of the latter, the 1967 was outstanding and in the January 1980 issue of *Decanter* magazine I wrote that it was one of the two top wines of the entire Fair. Dry but rich and fruity later that year and early in 1980. A glowing note: luminous, fragrant in 1988, and equally delicious 12 years later: rosehip maturity, still some tannin. Last tasted April 2000. (MB)

1966 *'I was trying to make a wine with my approach, but was still under the influence of Bordeaux. I tried to move in a new direction, but it was a very difficult vintage. This wine is completely different from my 1964. It is not the biggest Musar, it is light in colour with good acidity, but it started to drink well 20 or 30 years ago. This is the year that Gaston was born.' (SH)*

Mature, light, copper-coloured with a thin rim. Very powerful nose: not nuanced but big and solid; very vegetal and like boiled cabbage. Quite like a Cabernet Franc. Certainly not dead or dying; holding up really well. Lovely sweetness, almost jammy on the palate but develops amazing flavours on the aftertaste. Chewy, silky, totally mouth-filling, deliciously long with great balance. Old but very firm. Love this wine. Surprisingly, in Michael Broadbent's *Vintage Wine* it was last tasted in 2000 and awarded three stars, but it has evolved very well since then. **19** (BB)

1965 *No vintage made this year. 'I missed this vintage; I don't know why. Oh yes – I got married to Tania this year!' (SH)*

1964 *'I had been in Bordeaux. I was 25. I prefer not to qualify the attitude of the French towards me, but no-one hurt me: they respected my oenological degree. However, I never took them a wine of my own. I was pretentious, my way! Was it as a reaction to this? I don't know, but 1964 was my wild wine.' (SH)*

Mature rim but not for its age: still tons of vibrant life in its red colour. Very little sign of ageing. Nose is very big, more like a mature Châteauneuf-du-Pape or other big, structured Rhône wine. Very concentrated and deep with autumnal bonfire ash. Spectacularly good, huge on the palate: damp decaying leaves; great acidity; volatility balanced by brown sugar flavours; delicious, sweet, long on the finish. Hardly any age. A great classic wine. **20** (BB)

1963 *No vintage made this year. 'Only a* cuvée *wine was produced and none of the Chateau Musar top label. We made a "Chateau" wine only when it deserved it. Some years it did not deserve it.' (SH)*

1962 *No vintage this year. 'Only a* cuvée *wine was made; none of the "Chateau" top label'. (SH)*

1961 *'I was still experimenting, still learning.' (SH)*

Slightly sludgy rust colour. Sweet, rich and thoroughly exotic. Old leather handbags? With some biscuit slurry (dunked, not-too-gingery ginger nuts?). Ethereal and lifted. Beautiful sweetness as it hits the palate then real freshness and super-clean, thoroughly mature aromas. Gloriously mature but not at all old. Does taste like a wine from a bygone age... Long and rich on the finish. Amazing to think that Cinsault can last this long. No variety imprints itself. Just a little chewiness on the end suggests a solid accompaniment would be wise. 1980–2028. (**18.5**) JR

1960 *'That summer I was invited to Château Langoa-Barton for an internship with Ronald Barton. What I saw, tasted and learned there affected me. If I had to describe myself, I was a civil engineer who made wine and who had started to discover the complexity of wine. I was trying to understand, to observe. I knew that nature will never repeat herself and for a while I thought I might change all the grape varieties that we were growing, but to plant* grapes and make good wine takes 20 years, and what if I was wrong? However, I experimented a little with Grenache, Mourvèdre and Merlot. Looking back there were three stages to my winemaking. Between 1959 and 1964 I was making wines without an objective, with what we had at hand. From 1960 to 1970, having studied in Bordeaux and been influenced by Saint-Julien and Saint-Estèphe, I now realize that I was trying to produce wines in that style. Since 1970, I have been producing my wines, my way.' Les vins du terroir Libanais.' (SH)

Great depth for a wine of this age. Still very solid red with a touch of maturity on the rim. Looks about 15 years old. Classic nose: mature Bordeaux in character; signs of age and aromas reminiscent of very old cellars in France; lovely autumnal smells and meaty redcurrant fruit. Perfection on palate: full fruit and a great mouthfeel, with sweetness, length, great texture and perfect evolution; a little spicy and hot on the finish. Like an old Bordeaux but with better acidity, then a lovely aftertaste that conjures up the bonfire smells of autumn country walks. At its peak now. One of the best Musars I've ever tasted. **20** (BB)

1959 *'The year that I was born, my way,' said Serge Hochar, of his introduction to winemaking. Michael Broadbent said of the 1959 red: 'In 1979: fine, rich, mature, delicious bouquet, sweet, soft,*

fabulous. 20 years later: orange-tinged; lovely old nose; still delicious.' (SK)

The oldest vintage presented by Serge Hochar in London in August 1979: fine, rich, mature; delicious bouquet; sweet, soft, fabulous. Twenty years later: orange-tinged; lovely old nose; still delicious. Last tasted at Musar in Beirut, Dec 1999. **Five Stars.** MB

1958, 1957 No longer available.

1956 *Made by Gaston Hochar senior; bottled by Serge in 1959.*

Good colour for its age; remarkably sound, slightly chocolatey nose; surprisingly sweet, delicious, excellent tannins and acidity. Totally delicious. (At dinner with the Hochar family, Serge, his brother Ronald, their wives and Gaston junior on our first day in Lebanon, Dec 1999 **Four stars.** Drink up! (MB)

1951 *One of the few wines produced by Gaston senior that have been tasted by his grandsons. (SK)*

In December 2017, a nose of forest floor followed by a palate of sweat ripe berries, layers of tobacco, coffee, liquorice with balanced acidity and exuberating freshness. Elegant, subtle and still evolving beautifully three hours after decanting. Surprised that it had no sediments at all, probably fined or filtered when Gaston Senior was still at the helm. (MH)

CHATEAU MUSAR – WHITE

A blend of two local varieties, Merwah and Obaideh. Both grapes are partly fermented in oak barriques (25% new each year) where they mature for a further nine months. They are then blended, bottled and aged for a further six to eight years before release.

2012 Fine fresh-looking lemon-yellow, but shows reassuring maturity. Dry honey and spices on the nose; a controlled richness of fruit on the palate that reminds me of Ygrec: complex, dry but with sweetness in its DNA before natural acidity lifts the exotic aftertaste. Lanolin smooth but not easy to assess (and not meant to be at this stage). Young white Musar that retains its mystery. 2020–35. **18.5** (SS)

future. The next morning, the nose had become incredibly intense and showed enormous acidity. On the palate the acid was very powerful against the fruit, which was elegant and understated. By day three it had developed a fascinating, sweet, cheese aroma. **18** (BB)

2009 Fine light gold in colour, clear and fresh-looking in its 10th year. Yellow autumnal fruits, fresh herbs and dry honey on the nose, leading to a richly textured palate tempered by a line of acidity that creates a dry finish while retaining a blossoming richness that places it firmly into the *grand vin* category. At least another decade of evolution. To 2035. **18.25** (SS)

2011 Surprisingly full yellow with a faint orange tint from a late-picked ripe harvest. Freshness on the nose does not deny this, but a nutty clarity lifts the nicely concentrated, slightly exotic fruit and leads it to a dry finish. A wine that seems old for its years, on the way to resembling a dry Sauternes or an old-fashioned white Graves, with an intensity of fruit that will see it through into its next incarnation as a mature white Musar. To 2028. **17** (SS)

2010 Lemon-yellow colour. Lovely youthful florality on the nose with dry southern herbs. Fully flavoured on the palate for just 12%, showing the depth of fruit from old indigenous vines that were less affected than others in this heatwave year. Yellow summer fruits showing clarity and a fine lifted finish. Excellent now and will gain in complexity for another decade. **17.75** (SS)

Bright, clear yellow with no visible ageing. Very deep and intense on the nose with hints of sweet honey and balsa wood. Big, velvety and soft on the palate with great acidity, apple flavours and an edge of bitterness. Amazing development on the aftertaste indicating a long-lived wine with a great

2008 *An unusual year: snow in January and rain until February 23rd, then no rain for the rest of the growing season. The vineyards suffered. Harvesting was unusually early for the French varieties, but the indigenous varieties, Obaideh and Merwah, didn't ripen until October. (TS)*

Pale gold. Spicy (savoury, Indian spice) nose. I want to say ambergris – more out of onomatopoeia than aroma. Still rather sour without the roundness of mature fruit to counterbalance quite marked acidity which the accompanying notes attribute to the age of the vines. Definitely worth keeping until more complexity counterbalances the acidity. 2020–35. **16++** (JR)

2007 Fine, clear yellow, bright for 12 years old. The rich, slightly waxy, dry honeyed nose reminiscent of Sémillon shows no sweetness. A good breadth of fruit on the palate, lightly expressed intensity of dried fruit, richness of texture with honey and citrus elements present, then Mediterranean herbs over a persistent and lifted finish. Will continue to impress and gain more complexity for a decade and more. To 2032. **18** (SS)

2006 *'The 2006 white has a million lessons to teach us. It's already showing the infinite layers of aroma that can exist in a wine.' (GH)*

Medium-deep golden-yellow, bright and vibrant. A bit sleepy on the nose: subdued, creamy, with Petite Suisse cheese and Cheddar rind. A touch autumnal. Subdued and lean on the palate until the soft, creamy aftertaste brings it alive and indicates it will develop. Beautiful Musar but may put off anyone who is unfamiliar with the wine. Needs 20 years. The next morning, the acidity became brighter and more pronounced on the nose; fruit on the palate had gained complexity and intensity. By day three, the nose had gained even greater depth. **17** (BB)

2005 Mature yellow to light gold. Dry spices dominate the nose and natural richness is very present despite just 12% alcohol. Although dry, the palate exudes tropical fruits surrounded by a vigourously youthful acidity, each sip being more interesting than the last, with honeyed depth and unique old vines flavours for a fine future. To 2030. **18** (SS)

2004 Lively gold colour, still vibrant looking. Honeyed nose with a hint of orange peel that reveals a controlled richness of fruit and dry spices. Attractively leaner on the finish than expected and while very 'Mediterranean' it shows the profile of a finely matured white Graves. The natural acidity will allow more complexity to develop. To 2026. **17.5** (SS)

2003 Pale golden amber. I would never have guessed this was 14 years younger than the 1989 tasted immediately before it. Again, that lively edge of green vegginess on a broad palate with relatively low acidity and a lanolin quality that reminds me more of Sémillon than Chardonnay. Very sprightly on the palate. It isn't remarkably persistent, still has a little light tannin, and I can quite see – not least on the basis of the 1989 – that there is no hurry to drink this. Obviously oak aged with real structure. No more than medium body – but what a lot of flavour... Quite a challenge to work out the right food for this, I would have thought. 2015–30 **16.5+** (JR)

2002 *For the whites, 'there was only very small production – about a third of what we'd expect in an average year – due to a hailstorm on Mount Lebanon that obliterated the Merwah grapes. This is the only vintage that we haven't kept bottles from as one customer took almost everything.' (IS)*

2001 Full, deep-gold in colour, with fine clarity and a hint of amber. The nose shows a superb natural concentration of ripe summer fruits, and summer continues onto the palate with richness (not sweetness) amplified and balanced by a Tarte Tatin, quince-like acidity leading to a mellow, dry finish with endless length. An outstanding wine, first released in 2008 after six years in bottle to great acclaim. It still has a great future. To 2030. **19** (SS)

2000 Medium depth golden-yellow; no maturity showing. Still very closed on the nose, subdued, with a light floral nettle aroma. A bit astringent on the palate, but with mellow, soft textures and good acidity too. Needs time. By the next morning, the nose had come alive. The palate was beautiful, like a spicy Petite Suisse cheese, with great length. Nose continuing to develop on day three. **18** (BB)

1999 Full gold with orange tints, burnished and bright. An extraordinary bouquet blends pine cones, apricots, orange zest and toasted almonds. Full and richly flavoured on the palate and probably showed some natural sweetness when released in 2006, but the acidity has absorbed this to reveal a strikingly dry finish with inherent richness at the core. At 20 years old, this superb wine is in a Lebanese world of its own. To 2025. **18.5** (SS)

1998 Bright burnished gold, a superb and arresting colour in its 21st year. Rich, like Premier Cru Classé Sauternes on the nose, with ripe apricots and caramelized oranges, but not sweet. Deep yet elegant, fully rounded on the palate, with exotic fruits and dry marzipan finish. Marvellously textured flavours encased in a firm structure that stays on the palate and on the mind. To 2028. **18.5** (SS)

1997 Deep beautiful gold, of medium intensity. Nose deep and mellow, almost like a dry Sauternes. Good acidity. Beginning to be ready for drinking. Good length and a smooth aftertaste. Still needs time. The depth on the nose persisted into the next morning with a slightly walnuty aroma. On the palate, it developed the flavours of a very good fine cognac (without the alcohol). By day three the nose became distinctly sweet and fruity. 17 (BB)

1996 No longer available.

1995, 1994 Minimal stock available.

1993 *'A friend served a Château Haut Brion 1993 alongside my 1993 white. At first, my wine did not show. Half an hour later, she started to show herself. An hour later, the Haut Brion had faded, as my wine got stronger and stronger.'* (SH)

1992 *'The weather was terrible. We had winter until June. Hail came. Biblical weather! I tried to make a Chateau Musar red, but two years later found it wasn't up to standard, so we distilled it. The best grape that year was white Obaideh.'* (SH)

Deep orange gold. On the nose, beginning to mature and open up. Good depth and concentration. Lovely flavours of old Cheddar and surprisingly high acidity. Very good length on the aftertaste. Will age for a long time. After an hour in the glass, the wine started to develop characteristic honey aromas that also revealed themselves on the palate alongside a beautiful vanilla and beeswax aroma. The next morning saw it develop a rich, sweet, cognac-like nose; bone dry, powerful yet elegant on the palate with pronounced alcohol and a long finish. By day three, the middle nose had developed even more. **18.5** (BB)

1991 Very deep brown brassy-gold; age showing on rim with a little bit of amber. An intensely powerful nose, with sweetness, Ryvita, honey and jam notes and mouthwatering acidity. Bone dry, intense and spicy on the palate with high acidity, brown sugar and meringue flavours and great length. By day two it had become more spirity. **18.5** (BB)

1990 Bright, lively deep gold. Very attractive, youthful and lively on the nose, with elegant honeyed fruit, orange flower water and floral notes. More approachable than most white Musars. On the palate, very dry, with great acidity, very full fruit, length and spice on the finish. By day two had developed serious depth. **19** (BB)

1989 *A commemorative bottle to honour the legacy of Serge Hochar on his 50th anniversary.* (SK)

Pale amber wine with a very broad bouquet that hits you like a cloud not an arrow. This is not oxidized but is waxy, with a green, vegetal sort of Sémillon note on the nose. Then on the palate it has depth, real presence and something definitely salty about it – but citrus, too. It's certainly not rich in texture. Bone dry, with a hint of walnuts, and really quite light in the mouth. Truly unlike any other wine I can think of. It is indeed difficult not to find the idiosyncratic, impish but utterly assured spirit of Serge Hochar in this one-off wine. Fully mature but definitely not decayed. Extremely clean in fact. 2000–24. **17** (JR)

1988, 1987 No longer available.

1986 Bright yellow, pronounced, very youthful; no signs of age. Very young and vibrant on the nose with very fine fruit. Clearly defined. Syrupy but dry on the palate: this is an amazingly big wine with a huge concentration of fruit and great length. Spectacular depth and lovely fresh wine flavours. No sign of age: could not be considered any more than six years of age if tasted blind. An amazing, astonishing wine. Day two saw it become very serious and austere on the nose with flavours of vanilla and oaky herbal notes. **20** (BB)

1981 Bouquet of butterscotch, marshmallows, cooked apple and lemon. Dry and full-bodied with tart acidity. Very Granny Smith apple-like on palate, with mineral notes. Very long, with a toasty caramel finish (Luiz Gutiérrez). 1986–2000 **16** (JR)

1980 Minimal stock available.

1975 Deep orangey gold; looks like a late harvest wine in colour. Nose; wow! Incredibly intense, concentrated, dry, very deep and austere: a combination of leather saddles, barnyard and old mahogany musky characters with mouthwatering citrus orange acidity. Astonishing complexity and intensity on the palate, with a syrupy texture. Massive, deep powerful, concentrated and long. Delicious flavours keep changing like a kaleidoscope in this mouthfilling but elegant wine. On day two it had aromas of sweet Australian 'Tokay' and a bite of cognac-like alcohol. **20** (BB)

1972 Golden brass-yellow, very bright and alive. A superb nose with good depth and concentration, combining notes of apricot, brown sugar, cigar box and old charred wood. Unique. On the palate: a delicious, dry lean wine with a sweet finish and tremendous length. In perfect condition with a lovely lingering aftertaste. Very cerebral. Tastes of very old

cellars. Day two saw an amazing development of the nose: very hard to describe but utterly riveting to the brain. **20** (BB)

1970 Minimal stock available.

1969 *'This white had a sweet nose at first that is fading into something more... religious (heaven save us from my overwrought notes). After two hours, it's taken a more austere direction, but it is so long. A wine that lingers. I feel its glorious acidity in the corners of my cheeks. This is my monk's wine.'* (SH)

1967 Deep caramel colour. Dry, full-bodied with marked acidity. Vegetal character, with cooked apple, nuts and hay (Luiz Gutiérrez). 1975–2000. **16** (JR)

1966 Slightly dull yellow gold with brown hints; showing its age. Great complexity of old, very animally aromas; deep and concentrated with honey, cedar and good acidity. Definitely alive but showing great maturity. Palate: harmonious, stable, with an uplift on the aftertaste; sweet fruit on a dry wine. Amazing length. Past its prime but will show elegance for a long time to come. **17.5** (BB)

1964 Pronounced nose of caramel, candied fruit, pineapple and floral notes alongside beetroot, banana, pineapple and pear complexity. Dry, full bodied, pronounced and remarkably fresh on palate; piercing acidity and long finish. 1969–2013 **17** (JR)

1961, 1959 Minimal stock available.

1954 *An intense, raging white wine that made it into the 100 Vintage Treasures of French collector Michel-Jack Chasseuil. Produced by Gaston Hochar but bottled by Serge (SK).*

Bright gold in colour, and prodigiously aromatic, running through a chain of allusions which seemed

I hate make-up. That is why my wines have everything, but no make-up. Make-up and cosmetic surgery – in wines and in humans – change equilibrium.

———✦———

Life is energy, and energy is vibration. Where there is matter, there is energy. Who talks about memory, talks about life. That's why I talk about the miracle, the mystery of life. I cannot understand! I have come across a wall due to the complexity of nature, which goes beyond understanding. I only understand that life goes beyond science. Life is a permanent accident. If you are lucky, life is a lucky accident! It has been my great luck to be allowed to make my wines, my way.

———✦———

In vino veritas: there is truth in wine, and wine is a companion that has never betrayed me. Respect it and it will repay you. Wine is my friend, companion, medicine, advisor and doctor.

———✦———

My wines must have three things...Truth: nothing fake. Length: they should linger in the mouth. And Life: they must taste vital, like the land they came from.

———✦———

You can say that wine is between God and the Devil. And that it is a certain kind of heaven. Wine is also a link between God and humanity. Wine is tolerance. As it talks to you, you understand it more. And understanding breeds tolerance.

It can be tempting for a winemaker to interfere with the life of his or her wines. It can be tempting to treat a wine with interventions that make it more stable, or better behaved. But the only intervention I trust is trust. I will not strip anything out of my wines. I know that most winemakers filter their wines, and I have nothing to say for or against them: I can only speak of my wines. For me, the organic matter, those billions of tiny particles in a bottle of my wine, that is the brain of the wine. That is what allows the wine to mature, to be intelligent. If you strip out this organic matter, then (in my opinion) you lobotomize the wine. The wine may be stable, but it will not evolve in an interesting way.

———— ❦ ————

I woke up this morning thinking what do I want to achieve now? But I went back to sleep, so I don't know what I decided.

———— ❦ ————

Sometimes my brain functions well, sometimes oddly. I've learnt not just to rely on thinking but also on my intuition. Reasoning plus intuition have guided me. Now, if I have a crazy idea I go ahead with it. If I go against my intuition, I regret it.

———— ❦ ————

What is love? Love is the basis of humanity. Without love there would be no humanity. Love is a million things and one thing. Oh, yes – I am in love with my wines, my relationship with them has been a love affair. This is what I have been doing all my adult life – being in love with my wines. Love includes all things, you see: my relationship with my wines is a give and take relationship. It is very good not to understand all the mysteries of life.

Appendix II

TRIBUTES TO SERGE

The wine world pays its respects

ANDREW JEFFORD, wine writer, France

FOR 27 YEARS, I'VE TRAVELLED and talked to wine producers around the world. These discussions are sometirne repetitive, insincere, tedious and superficial. More often, though, they are interesting, because wine producers are both farmers and craftsmen. They rummage in earth, and prune vines under capricious, cruel skies; they then press their harvest, and fashion humankind's most sensually alluring and nutritious drug with it. The fact that they sell not raw produce but a crafted product to an international market requires them to be articulate, and their journeys to meet those customers give them un-parochial perspectives.

Never, though, did I meet a winegrower who said more interesting things to me than Serge Hochar. I didn't agree with all of them but nothing he ever said left me unmoved.

He told me that he wanted to die when he was 15, though he was not depressed. From the age of seven, though, he had sought logic in all things, to the extent that his family called him 'Logic Serge'. He could find no logic to his existence. 'I couldn't understand why I was alive.' The question was shelved for some years, and Serge's existence continued, pending a response. Eventually he found one. 'Wine was a very satisfactory answer to my question. It was indirectly an answer to the mystery of existence. It made me accept life. Wine is not human life, but it is real life all the same.'

We never stopped talking about taste. Latterly, he came to believe that much of wine's significance lay in its 'taste beyond taste', by which he meant that the physical responses provoked in your body by the taste of wine are much less significant than the mental, emotional or spiritual responses it might provoke in your brain. He had no time for point scores ('You look at a girl and you like her. You look at another girl and you don't like her. But your friend likes the girl you don't like'). Customary tasting notes, with their daunting adjectival burden, are barely more helpful when it comes to tracking the 'taste beyond taste'. For that, you have to sit with a bottle of Musar and give it time to murmur in your synapses. 'Give it more time,' many heard him say, 'and my wine will give you more pleasure'. Thank you, Serge: we found that to be true.

GEORGE KANAAN, Chairman, Arab Bankers Association

IT MUST HAVE BEEN 1994 and Serge and I were eating at The Square in London. When they discovered who Serge was, they treated him like a king! The old sommelier decided to test Serge and brought out an unlabelled half bottle with dessert. Serge smelled, tasted and pronounced the château, the year, even the name of the winemaker. Then he stopped: 'No it's not possible. There's something different about the style. A nuance, but different. There's an echo of the influence of...' and he named a winemaker.

I thought the sommelier was going to cry. 'Monsieur Hochar,' he cried out. 'Nobody has a palate this fine. It's true, the château very quietly changed winemakers. The man you say it is made this wine, but as a tribute to the style of his predecessor.'

LAUREN FRIEL, wine writer, Massachusetts

I ADMIT, THE FIRST TIME I met him five years ago as a young and cautious sommelier, I wasn't sure what to make of this man who spoke about wine as most men speak about religion. But as I've grown, I've come to understand that the men and women who speak in such a way about their work are the true artists of this world, whether they are painters, musicians or winemakers, these people are the people who are here to show us what is beautiful about this life. We remember all of the ways in which Serge helped us see the beauty in the world. It is a gift that has no boundary.

MICHAEL KARAM, wine writer, *Harpers*

ULTIMATELY, SERGE HOCHAR was a fighter; for wine, which he believed unlocked the door to human understanding; for his country, which he saw as a glorious, magical and even mystical land, the crucible of all things human and good, and for what it meant to be Lebanese, to be blessed and cursed in equal measure, but to trust in the ultimate triumph of human spirit.

STEVEN SPURRIER

IN THE LATE 1970S, Serge visited my wine shop in Paris to introduce me to Chateau Musar. I began to stock the wine and thus began a friendship that I never thought would end. Serge represented the art of the possible wrapped up in a cultured, almost old-fashioned, charm: a loveable inspiration to those whose life is wine. His passion for everything about wine, especially for Musar, was totally modern. Of course I met the families, for although Serge was the 'face' of Musar, for him it was a family business.

JANCIS ROBINSON MW

SERGE WAS SO MUCH MORE than a wine-maker and the driving force behind Lebanon's best-known winery. He had a strong spiritual character, but was very far from ascetic – he was positively impish in fact.

Always great fun, he gave the impression of having a deep understanding of human nature and of understanding much more than superficialities. We talked about very much more than wine.

ADAM MONTEFIORE, wine writer

I WAS SO SORRY to hear of the untimely death of Serge Hochar, a true cedar of Lebanon. He was an elegant, charming figure, in a pinstripe suit, with all the exaggerated hand movements to support his French-accented English. It was the mischievous twinkle in his eye, gap-toothed smile and wonderfully expressive eyebrows that caught your attention.

He was a wine philosopher who would answer the most basic question with one of his own. An example was his querying the reason for wine's place in the universe. 'I know nothing about wine,' he would say, 'and every day I discover I know less'.

Chateau Musar is a unique expression of the Hochar *laissez faire* mode of winemaking. The enigmatic Musar white is even more unconventional. In a world of globalization and technical perfection, Serge not only insisted on doing things in an eccentric way, he delighted in being different. Wine today reeks of sameness. Musar is the ultimate expression of individuality, originality and distinct sense of place.

The Psalms say: 'The righteous shall grow like a Cedar of Lebanon.' I feel privileged to have met him. Thank you, Serge, for the memories and the inspiration. This cedar has fallen, but Musar continues.

BARTHOLOMEW BROADBENT, founder and chairman, Broadbent Selections Inc.

SERGE HOCHAR WAS PROBABLY the most respected, admired and influential winemaker I have ever met. He was beyond being a winemaker; he was a philosopher who managed to connect the dots between wine and civilization, and wine and life. He was inspirational and had a deeper understanding of the importance of wine than any other person alive.

Serge influenced generations of natural winemakers. His wine defined natural and organic winemaking, not as a trend-setter but as an interpretation of what wine can truly be.

He was the first *Decanter* magazine 'Man of the Year' in 1984 – an award that was created for him but subsequently given annually to the greatest and most influential of wine luminaries. He was awarded this ultimate recognition not only for making world class wine but also because he endured greater hardship than almost any other winemaker. He successfully made wines in every year but one of the Lebanese civil war. The year he failed was not for lack of trying, but because his vineyards became the centre of a battle ground.

When Serge spoke, the entire wine world listened. Whether you liked his wine or not, you respected him. He was humble, he listened and he coined the most deliciously quotable phrases, which we now call 'Serge-isms'.

Just as the vine develops to be an integral part of the land, producing wines of depth and intensity as it ages, so Serge came from a culture that has been making wine since the dawn of civilization and from this gained an understanding that could never be learned at any winemaking school.

As his teacher, Professor Emile Peynaud, said to Serge: 'I can teach you knowledge but I cannot teach you know-how.' Serge learned to know how, and became the most universally influential individual the wine world has ever known. My father, Michael Broadbent, was more saddened by his untimely death than almost any other in his lifetime. It is because Serge was everything and everywhere. He was the Pope of wine.

MICHAEL BROADBENT

HE WAS ONE OF MY dearest friends. He was a hero, making the most extraordinary wine with shells whizzing over his head. Generous and lovable, he was an amazing man, and an extremely good winemaker. He was one of the great characters, and he will be missed.

OZ CLARKE, TV presenter and wine writer

SERGE AND I NEVER talked about winemaking. I never knew what types of barrel he used, or whether they were new or old. I never learnt how long he macerated his grapes for, or whether he pumped over, or punched down, or racked, or returned. No. I'm sure he knew, but whenever I managed to find him he wanted to talk about more important things – poetry, philosophy, his beloved, troubled country Lebanon – oh, and wine. But not any wine. The wine which he would copiously pour into big glasses for any who was near. His wine. And then he would toast us, drink deep – I never saw him spit, so nor did I – and off he would go, like an impish, playful, rogue-eyed middle-eastern elder holding forth between long draughts on his hookah.

Chateau Musar needs more than a mouthful to understand it. More than a bottle, maybe several bottles, because it always changes, and always for the better. 'That's the spirit and soul of the vintage', Serge would say, the flavours twisting and turning in your mouth, in your mind, leading you up blind alleys, leading you towards false dawns, but always finding a final, fascinating, frequently paradoxical expression of this vineyard and the grapes it ripened in this month of that year.

Serge said that he always blended by following the deeply emotional and personal perception he had of each particular vintage. And he never knowingly tried to make a safe wine, a wine that could be guaranteed a respectable critical reception.

'My favourites are the ones which are animal, which are wild', he said. 'What I want is a wine that troubles me. Great wine should be dangerously attractive rather than simply enjoyable.'

Serge needn't have made his wine in Lebanon. He had trained as a winemaker in Bordeaux. For 10 years he then searched for a wine estate far from the fractious Levant. But he couldn't stay away from the country that was his passion, his dangerous passion, his wild passion. So he returned, and became the blazing beacon for Lebanon and its wine around the globe, for the oldest, the truest reason in the world.... because Lebanon was his home.

CHATEAU MUSAR, by Edward Ragg MW

'We are inoculated to resist.'

SERGE HOCHAR

The shade of the pergola,
the safest of vineyards, brings
no ripeness to this land.

To be rooted here
is to squat low
yet remain upstanding,

to know that each
juice-taut grape, each
cluster bunch held

is as permeable as skin
yet more symbolic,
that resilience has

no temperature,
survival no sun
except the oldest one.

First published in *A Force That Takes*
(Cinnamon Press, 2013)

Index

Page numbers in *italic* refer to illustrations

Acknowledgements

Ronald, Gaston, Marc and Ralph Hochar would very much like to thank the following for the loyal and unwavering support they have given Chateau Musar over the years.

The late John Avery and the Avery family (UK)

The late Gerard Basset OBE MS MW

Jeanne Beth (Hong Kong)

Michel Bouverat (Singapore)

Bartholomew Broadbent and his entire Broadbent
   Selections team (USA)

Michael Broadbent (UK)

The late Hugo Broen (Norway)

Maggie Chan (Hong Kong)

Michel Dovaz (France)

Jurgen Drawert (Germany)

Tim Evans (Australia)

Philippe Faure-Brac (France)

Roland Fischer (Switzerland)

The Gargano family (Italy)

Geir Gjerdrum (Norway)

Paul Grieco (USA)

Luc Hoornaert (Belgium)

Catherine Miles (USA)

Jim Nicholson (Northern Ireland)

Arvid Nordquist (Sweden)

Martin Paulsen (Norway)

Sophie Riby (Italy)

Jack Segal (Canada)

The late Derek Smedley MW

Steven Spurrier (UK)

Henrico van Lammeren (Netherlands)

Jan van Lissum (Netherlands)

James Tidwell (USA)

Michael Whiteside (UK)

Anette Wyren (Club Musar, Sweden)

Oliver Yan (People's Republic of China)

At the winery

Tarek Sakr

Charbel Abi-Ghanem

Tony Ammar

Fadia Kadamany

Tamara Ammar

At the Chateau Musar Beirut office

Desirée Khoury

Fadi Kalaani

Gebran Birack

Sossi Charlakian

Camille Ghannam

Georges Khoury

Gisele Helou

Rita Hrairy Rabbath

Tony Roukoz

Adel Kaed

Danny Kaed

At Chateau Musar UK

Jane Sowter

Richard Hunt

Camilla Shepherd

Jane Thomson

Elliott O'Mara

Naomi Hodges

Picture credits

FRONT/BACK COVER AND ALL INSIDE PHOTOGRAPHY BY LUCY POPE except: endpapers Chateau Musar; p6 The Hochar family; p9 theogould.com; p10 Cephas Picture Library. **Chapter 1:** p15 Google maps; p17 The Hochar family; 18 The Hochar family. **Chapter 2:** p27 The Hochar family; p29 (l) JancisRobinson.com, (c) Christie's, (r) Getty Images; p32 Shutterstock. **Chapter 3:** p 38-39 Chateau Musar; p40 Google maps; p42 Chateau Musar. **Chapter 4:** p57 Shutterstock; p64 Chateau Musar; p68 Chateau Musar. **Chapter 5:** p76 Alamy Images; p78 Alamy Images; p79 Metropolitan Museum of Art; p79 Getty Images; 80 Alamy Images. **Chapter 6:** p93 The Hochar family/P-D Art; p95 Getty Images; p96 Getty Images p98 Getty Images; p101 Getty Images. **Chapter 7:** p104-105 The Hochar family; p107 P-D Art/Erik Charlton; p109 The Hochar family; The Hochar family; The Hochar family; p113 The Hochar family; p116 The Hochar family; p117 The Hochar family; 118 Getty Images; p120 Alamy; p121 The Hochar family. **Chapter 8:** p124-125 Alamy images; p126 Hochar family; p127 (all) The Wine Society; p128 (l) courtesy of West Knowle Media Centre, (r) Christie's; courtesy of Allan Benoit/deepix.com; p130 Hochar family; p131 Chateau Musar; p133 Sarah Kemp; p137 Paul Moore; p138 Bartholomew Broadbent; p139 Paul Moore; p140 The Hochar family; p141 P-D art; p142 Getty images; p143 Dragon Phoenix Wine Consulting; p143 Dragon Phoenix Wine Consulting; p146 Alamy Images. **Chapter 9:** 148-149 The Hochar family; p160 The Hochar family; p163 Chateau Musar. **Chapter 10:** p169 (l) JancisRobinson.com (l2) Christie's; p185 The Hochar family. **Appendices:** p190 Andrew Jefford; p191 George Kanaan; p192 (t) Tom Fletcher, (b) Sarah Kemp; p193 (t) Fiona Beckett, (b) Jeremy Parzen; p194 (t) Lauren Friel/Ella Rinaldo, (b) Michael Karam; p195 (b) Jancis Robinson.com; p196 (t) Svein Vinofil Lindin, (b) Luca Gargano; p197 Lucy Hope; p198 (t) Adam Montefiore; p199 Cephas Picture Library; p201 The Hochar family.